Love and Instinct

Glenn Wilson

Routledge
Taylor & Francis Group

First published in 1981
by Maurice Temple Smith Ltd

This edition first published in 2014 by Routledge
2 Park Square, Milton Park, Abingdon, Oxon, OX14 4RN

Simultaneously published in the USA and Canada
by Routledge
711 Third Avenue, New York, NY 10017

Routledge is an imprint of the Taylor & Francis Group, an informa business

Publisher's Note
The publisher has gone to great lengths to ensure the quality of this reprint but points out that some imperfections in the original copies may be apparent.

Disclaimer
The publisher has made every effort to trace copyright holders and welcomes correspondence from those they have been unable to contact.

ISBN 13: 978-0-415-72331-2 (hbk)
ISBN 13: 978-1-315-85164-8 (ebk)

GLENN WILSON

Love and Instinct

London

First published in Great Britain in 1981
by Maurice Temple Smith Ltd
37 Great Russell Street, London WC1

ISBN 0 85117 209 1

Photoset in 11/13 Garamond by
Robcroft Ltd, London WC1

Printed and bound in Great Britain
at The Blackwell Press,
London, Guildford, Worcester and Oxford

Contents

The final aim of all love intrigues, be they comic or tragic, is really of more importance than all other ends in human life. What it all turns upon is nothing less than the composition of the next generation. *Schopenhauer, 1871*

To Dorothy and Daniel
with my love

1

Introduction

When Charles Darwin put forward his theory of evolution in the middle of the last century, it was greeted not just with scepticism, but with emotional resistance. It seemed to threaten people's self-esteem, based on the view of themselves as a unique and special creation. The wife of Bishop Wilberforce was most dismayed at the news of Darwin's blasphemy. 'Descended from monkeys?' she exclaimed. 'My dear, let us hope that it isn't true. But if it is true, let us hope that it doesn't become widely known.'

This ostrich-like posture is seldom encountered today in such unsubtle form. Most of us have accepted the fact of our evolutionary origins, at least with respect to our anatomy and physiological functions. But it still comes hard with many people to acknowledge that our thoughts, feelings and social behaviour are also powerfully influenced by an animal heritage – the struggles of our remote, prehistoric ancestors to survive and reproduce themselves successfully in hostile and competitive environments. We have a strong inclination to regard our own behaviour as following a path of 'free will' or rational choice, in contrast to the brutes, who are pushed and pulled by compulsive survival instincts.

This is a misleading conceit. No doubt the animals we think of as inferior near-automatons also have an illusion of free will and think of themselves as making reasonable choices, if only we could ask them. If we can bring ourselves, at least temporarily, to put aside understandable pride in our superiority, we may be surprised at the extent to which the complexities of our behaviour can be understood as arising out of instincts which at some time in history have conferred an evolutionary advantage upon us. This book deals with the application of such evolutionary explanations to human sexual and romantic behaviour.

The concept of instinct is admittedly unfashionable in modern psychology. It initially fell into disrepute because it sometimes gives an illusion of explanation where none in fact exists – as when we attribute the sociability of people to their 'gregarious instinct' or wars to the instinct of aggression. Of course such statements are circular, but the idea of social learning has been used in an equally sterile way. Explanations of social behaviour in terms of modelling our activities upon patterns prevalent in the culture are just as simple-minded; they beg the question as to how and why these behaviour patterns were adopted in the first place.

Another reason why instinct-based theories of social behaviour have lost popularity in recent years is to be found in the work of the cultural anthropologists. These researchers have tended to observe and document differences between one cultural group and another, while ignoring many important uniformities. This is perfectly understandable because the differences are a great deal more interesting than the similarities. The process is akin to that of the tourist who selectively recounts those aspects of a foreign country which are different from what he is accustomed to at home. There is no reason why he should be impressed by the things that are the same.

The result is that anthropologists have, over the years, led us to believe that human social behaviour is very largely 'culturally determined'. There are, they maintain, very few biological imperatives beyond the most obvious physiological functions like breathing and sleeping. For almost any type of behaviour that we think of as natural, the anthropologists have come up with some obscure cultural group that does not conform to the usual pattern, and the conclusion they have reached is that our behaviour is virtually free from instinctual control.

Yet cultural exceptions do not really prove the absence of an instinct. As Desmond Morris notes in *Manwatching*, nuns and priests supposedly live out non-sexual lives, but this should not be taken to mean that sexual intercourse is a non-biological, cultural invention on the part of the rest of us. Exceptional communities do not disprove any rules, they merely illustrate the range of social control and modification of our instincts that is possible.

Although almost taboo in some branches of psychology, the instinct concept is by no means lost and forgotten. Psychoanalysts, as well as most laymen, regard it as inescapable that much of our behaviour is motivated by primitive, often unconscious, biological forces which burst through our civilised veneer on occasions with irrepressible force (sometimes to our guilt and embarrassment).

There is also a thriving branch of biology, called ethology, which is in effect concerned with the study of instincts. Recently, interest has focussed on the explanation of complex social behaviour such as altruism and courtship in terms of optimal strategies for survival. A conceptual breakthrough occurred when it was recognised that the unit of natural selection is not the individual animal but the individual genes, and this was the basis for the whole new discipline of 'sociobiology'. It is my belief that social psychologists can no longer afford to ignore the insights produced by these approaches.

An example may show how the evolutionary perspective can sharpen the examination of social behaviour. It has often been supposed that animals (particularly female animals) have an instinctual need to produce offspring, and that such an instinct would inevitably confound efforts within human society to implement programmes for population control. Yet if we assume that instincts have to evolve just like anatomical structures, it is in fact very unlikely that we have any drive to reproduce or have children *per se*. Non-human animals have no awareness of the link between copulation and reproduction, so even if they wanted to have offspring they would not know how to go about it. The same would apply to certain primitive human societies who, until recently at least, had not discovered the copulation-reproduction connection.

For effective reproduction, nature required the development of only two biological drives: (1) a sex drive that would lead male animals to copulate with the females, and (2) a parental care instinct that would motivate adult members of a species to protect their offspring once they have been born. There is no need for a bridging instinct of any kind, or indeed any means by which animals with such an inclination could be selected for. The sex drive and the instinct of parental care can operate quite independently, and have a separate neurological basis, yet both serve the evolutionary end of reproduction.

Of course, in modern human society the position has changed. We now have a fairly full awareness of the natural consequences of sexual intercourse and we have effective means for preventing them (contraception and abortion techniques). Therefore, conditions have been established which *would* allow for the evolution of a new instinct to bridge the gap between sex and child-rearing – an instinct that might be called 'child-acquisition'. But while this new drive may be built into human nature many generations hence (as a result of the breeding advantage of people who desire children) there has not been sufficient time for this social change to have filtered through to the level of our genes.

At the present moment it is most unlikely that humans have any need to acquire children, at least not one that is deeply rooted in biological instinct. There may be social learning experiences that would lead young couples to think it might be nice to have children, but it would be equally possible for experience to operate so as to deter couples from starting or adding to a family.

And if there is no biological instinct for acquiring offspring beyond a simple sex drive, what will happen to the human species once we are perfectly able to separate recreative from procreative sexual activity? It would be no surprise if the much feared population explosion were to give way to a population 'implosion'. Indeed, there are signs that this is already happening in many advanced Western societies.

This example has been discussed in some detail because it shows how biological instincts may be distinguished from learned needs, and the social implications that may be drawn from that distinction. Instincts can only evolve if there is some logical way in which a feeling can be translated into a behavioural pattern that will confer some selective advantage. By studying the conditions prevailing at an earlier stage of our evolutionary development it is sometimes possible to deduce that a particular aspect of human behaviour could not readily have arisen out of biological forces.

A case in point that will be discussed more fully later in the book, is that of female ability to reach orgasm. Since the females among our mammalian and primate ancestors do not seem to have experienced orgasm at all, it is difficult to see how it could have been selected for. And if there was an evolutionary advantage to the human female orgasm, then it is hard to understand why its appearance should be so variable and elusive. Later in the book it will be argued that the human female is not biologically distinguishable in this respect from any other primate. They will all achieve orgasm if given a sufficiently long period of suitable clitoral stimulation but there is no particular biological advantage in their doing so. Female orgasm appears to be a socially learned pleasure, like playing the piano or smoking marijuana. However enjoyable, it is not, in the biological sense, a 'natural' function.

An almost opposite state of affairs is that of the loss of a marked period of sexual receptivity or 'heat' around the time of ovulation in the human female. Throughout the evolutionary chain there seems to be a progressive relaxation in the grip that ovulation has on female sexual copulation, so that in the human female it is difficult to detect any cyclic fluctuations in libido at all. The so-called 'rhythm method' of birth control is of course notoriously unreliable, and some theorists have suggested that its ineffectiveness might be due to powerful female sex urges emerging at the time of peak fertility. But the fact is that, despite concerted attempts to do so, no one has yet produced any unequivocal evidence that women are driven to copulate at the point of ovulation. There is evidence that men find women more attractive at this time of the month than others, but no real reason to suppose that women's defences are down.

At first sight this seems paradoxical from the evolutionary point of view. Copulation in mid-cycle ought to be more productive (that is, reproductive) and should therefore have a selective advantage. The only way to resolve this paradox is to think of some way in which a powerful peak of female libido at ovulation might be *dis*advantageous to an advanced primate. Fortunately, such a drawback is readily apparent. If a woman were to 'lose control' every time she became fertile, she would no longer be in a position to choose carefully which man was to be the father of her child and at what point in her life conditions were most favourable to child-rearing. It is this need to retain a degree of rational control over her sexual favours that has led to progressive emancipation of the higher primate females from the cyclic control of their receptivity. Primate females are able to select males of superior breeding stock to sire their offspring and become impregnated at times when they are well-equipped for successful parenthood (for example, when they are assured of food, protection and other forms of support).

Interestingly, these concerns that we recognize as being characteristic of the human female are usually thought to be socially learned. It is true that they are taught, but evolutionary analysis suggests that, beyond that, female caution and control has evolved in higher primates in the form of a biological instinct. Social factors alone would not be enough to account for its slow, progressive development from one species to another.

Similar control over the sexual urge is of course less relevant to male primates, since getting the 'wrong' female impregnated does not preclude getting the 'right' one pregnant later on. It may even be to the genetic advantage of males to distribute their seed widely, even if some of it falls on stony ground. Females only have a certain number of eggs in their basket and need to use them very carefully in order that their genes will survive in the next generation. Males have virtually unlimited opportunities for reproduction and so can afford to scatter their shot. This is called 'parental investment theory' and it explains a great deal about human courtship and mating behaviour that previously seemed very mysterious.

Consider the eternal sex war. Men and women have been aware of certain incompatibilities of thought and feeling throughout history, at least since the days of Lysistrata. There has never been a time, as far as we know, when men and women have seen exactly eye to eye, and the present is certainly no exception. A commonly expressed feeling about women in male circles is that 'you can't live with them, and you can't live without them'. No doubt a high proportion of women feel just the same way about men. Men find women too often indecisive, unadventurous and clinging, while women frequently find men over-aggressive, irresponsible and unfaithful. Each sex may appreciate certain virtues in the other, but the differences between them seem almost inevitably to lead to conflict in the course of an extended relationship.

It is becoming increasingly fashionable at the moment to regard these gender differences as unnecessary effects of social training which could be eliminated if little boys and little girls were raised identically. Some people see men as having acquired their aggressiveness and lustfulness through social learning, and believe that in an ideal environment they would become civilized, like women. Others see women as being rather like repressed men, who given sufficient opportunity would prove to be just as assertive and sexually playful as men are today.

A central theme of this book is that men and women are by no means the same animal dressed up differently and that an evolutionary study of their separate development explains why. There are very good biological reasons why men and women should think and feel differently from one another, just as there are good reasons why they should have become physically differentiated in the way that they have. In fact, the mental and emotional differences between the sexes may be very much greater than most of us have previously supposed. For diplomatic and strategic reasons men and women throughout history may have pretended to share more of the motives and ideals of the opposite sex than they really do. This argument will be taken up particularly in Chapters 2 and 3.

On the other hand, it is important to bear in mind that there is a considerable overlap between the sexes in all traits, whether physical, mental or emotional. Throughout the book we shall be discussing differences between average or typical men and women, but that is not to deny that there will be many exceptional individuals to whom the rules do not apply. No doubt those most atypical of their gender, especially when surrounded by like-minded friends, will find the evidence for instinctual differences between men and women most difficult to credit.

It was earlier stated that instincts became unfashionable in psychology because of certain scientific difficulties with the concept. But there is another reason why they have not readily been restored to favour. The current *zeitgeist* in social science favours environmental explanations of behaviour because they are politically and philosophically more comforting. They fit better with the egalitarian and self-help attitudes which are prevalent in the modern world, especially the United States. There is a fear that biological theories of social behaviour will engender a kind of fatalism – a feeling that we are unable to escape or supercede our animal origins. There is even a tendency to confuse explanation with justification. If it is 'natural' for men to be more promiscuous than women, for example, this might seem to imply that sexual control is neither possible nor desirable for them, that the double standard is normal and inevitable.

This fear is quite understandable, for there is an

element of truth in it. Humans are partly the victims of their biological instincts, but not entirely. Ethical considerations, arising from the needs of social living, also have power to modify our behaviour. If it is possible for biological theories of the kind discussed here to be invoked in such a way as to diminish the effectiveness of ethical injunctions, it is also a truism that any scientific knowledge can be applied for good or ill – just as atomic fission can be used to generate power or destroy cities. The truth is that brutal instincts are a significant factor in our mental and emotional economy but cognitive and moral control is usually still possible. Promoting recognition of the true power and role of instincts is not the same as advocating the total abandonment of social restraint. It is better to know the Devil than pretend he does not exist.

2

Male and Female

All of what we call sexual and romantic behaviour occurs because, roughly speaking, we are divided into two sexes. Without gender differentiation none of this pantomime would be necessary. We would not feel that 'incompleteness' which moves us to seek long or short term partnerships. A discussion of the origins and reasons for sexual differentiation is essential for an understanding of the instinctual basis of sexual behaviour.

Why two sexes?

Why is it that the vast majority of animals (and plants) are arranged as two sexes? Would it not be a great deal easier just to split ourselves in half like amoebas? Most people take this gender division for granted, or if moved to think about it at all, will usually offer some pseudo-Biblical explanation like the idea of men and women needing each other as companions. This, of course, is a circular argument, since if we were complete in ourselves we would need no companion.

A satisfactory scientific answer to this question must involve evolutionary principles and in particular the concept of 'reproductive success'. All living creatures survive through history and gain ascendancy in accordance with the extent to which they are successful at producing facsimiles of themselves that have a good chance of reaching reproductive maturity. Increasingly, though, it is becoming recognised that the real struggle for survival takes place at the level of the genes, which are the basic units of inheritance. Evolution is best understood if we think of genes as constructing machines that are optimally suited to carry and reproduce themselves. These mobile devices are what we identify as the adult animal. Now since several individuals may be carrying the same genes to some extent (especially if they are closely related family members) it makes sense for us to be concerned with the survival not just of ourselves but other people as well. Hence we find altruism occurring in direct proportion to the amount of genetic overlap between the helper and the helped. Asked whether he would sacrifice his life for his brother, the biologist J.B. Haldane replied no, but he would for three brothers or nine cousins. This, he calculated, would maximise the survival of his genes, and is similar to the instinctual rule-of-thumb applied by most animals when they come to contemplate acts of heroic self-sacrifice.

Now how does this theory of gene survival apply to the characteristic division of animals into two sexes? At first sight sexual reproduction would seem to be disadvantageous. After all, amoebas pass on one hundred per cent of their genes to their offspring, reproducing themselves perfectly, whereas we contribute only fifty per cent of our genes to our children (the other fifty per cent being contributed by the other parent, who usually is not closely related to us). Add to that the fact that our system is cumbersome, time-consuming and generally messy, and we might reasonably ask why we should have adopted it at all. Surely it must have some kind of advantage over simply splitting ourselves in half for it to have evolved in the first place.

The main drawback to non-sexual reproduction of the kind practised by amoebas is paradoxically connected with the very fact that all the individuals are genetically identical, like cloned frogs. This being the case, no adaptation to the environment can take place except by mutation or learning. Thus a major source of adaptation is unavailable. The process of sexual reproduction leads to genetic diversification, all of the individuals in the species being a little bit different from one another (with the partial exception of identical twins). This variability is the key to our adaptability. If conditions become difficult, some individuals will be better equipped to survive than others, and they will form the core of future generations. The situation has been likened to a lottery in which sexual reproducers hold tickets marked with different numbers while non-sexual reproducers hold several tickets all marked with the same number. Some animals, such as the barnacle and sea-bass, are capable of fertilizing themselves if isolated for long enough, but because of the advantages of reshuffling genes, they have an instinctual preference for mating with another of their species.

Now if two sexes are better than one, would not three sexes or even more be better still? It seems not. Because nearly every healthy individual mates with another genetically unique individual, this system provides enough potential for the reshuffling of genes. Two sexes are sufficient to produce maximum diversity in the offspring. This is probably just as well, because human relationships are difficult enough when the compatibility of only two people has to be considered. Imagine the domestic squabbles and divorce rate if we were required to live together intimately in threesomes, foursomes or moresomes, each of the several sexes having its own little foibles.

Sexual behaviour as the basis of species separation

The ability and willingness of male and female animals to interbreed is normally taken as the definition of species. Members of the same species practise 'reproductive isolation' from members of other species; they do not in the normal course of events mate with them. This frequently means that different species would be 1physiologically incapable of cross-breeding for one reason or another, but not necessarily so. Many species are apparently capable of cross-fertilizing but for reasons best known to themselves choose not to. Human interference has led to a number of hybrids which do not occur in nature, though sometimes the offspring are infertile, as in the well-known case of crossing a horse with a donkey to produce a mule. But the limits of cross-breeding are still being explored. Attempts are currently under way at the Yerkes Primate Research Centre in Georgia to produce chimpanzee/gorilla hybrids. There has even been discussion of the feasibility of crossing humans with apes. The main problem is that most apes have 48 chromosomes while human beings have only 46, and this would seem to be an almost insuperable problem. Recently, however, there were reports that a Japanese woman (a nineteen-year-old, out-of-work actress) had volunteered to attempt mating with a Congolese ape called Oliver, who had been discovered to be exceptional in having 47 chromosomes. Since the outcome of this experiment was never reported it presumably turned out to be a scientific failure, whatever it may have done for the lady's film career.

While it is true that the various species do not, in the wild, attempt sexual relationships one with another, the choice of one's own species as a sex target seems to be dependent on some kind of parental imprinting. As many a pet owner has found to his or her embarrassment, domestic dogs and cats are just as likely to get amorous with their owners as they are with others of their kind. Similarly, there are many recorded instances of zoo animals who cannot be induced to breed in captivity because they are more aroused by their keepers than the mates that have been imported for them at great expense and loss of life. From an evolutionary point of view it is perhaps strange that the preference for making love to members of one's own species is not more firmly rooted in instinct. Presumably it is unnecessary for the most part, since animals are normally surrounded by their parents, siblings and other 'conspecifics', at least for the early years of their life when the sexual object becomes established.

The sex ratio

The next question we might ask is why, in most species, there is an approximately equal number of males and females. (Actually, with humans there are slightly more male children born than female, but, as if to balance things up, more males than females are eliminated by disease, accidents, and so on.) Darwin himself confessed to being puzzled by this form of sexual equality, but later theorists have come up with a fairly logical argument (Fisher, 1930). Suppose one sex is under-represented in the population for some reason. Members of this sex will then, on average, leave more offspring each than members of the commoner sex (since each member of the rarer sex will have to mate more than once for each member of the commoner sex to conceive). As a result, any genes predisposing a parent to produce children of the rarer sex will increase in frequency until the numbers of each sex are again equal.

The chance factor in determining the sex of a child works at the level of the individual person, in fact, even the individual birth. It is sometimes believed that there are some women or couples who are prone to having children of one particular sex, the result being that they produce long runs of girls or boys in their family. However, research using large samples of people has shown that the sex of our previous children in no way predicts the sex of our subsequent children. Parents who have produced a long line of girls for example, are no more or less likely to produce another girl at their next attempt. The chances appear to be about 50/50 on each occasion regardless of the gender history of the family. It is, however, true that, to some extent, twinning runs in families, being carried particularly by the female line.

It is possible to influence, if not actually determine, the sex of your child by artificial means. Among the techniques that appear most promising are the testing of vaginal acidity, sperm separation with artificial insemination, and selective abortion. Thus, as with many other aspects of our evolution, we may soon be in a position to change the sex ratio at birth according to our preferences. Quite what effect this will have on the population is difficult to determine. Traditionally, it was supposed that more parents would choose to have little boys because they are economically more useful and carry more prestige in patriarchal societies. However, the preference for boys is much less marked in modern society. In fact, many parents today prefer little girls because they are more decorative and less rebellious than boys. Surveys of parental preferences indicate that, as things stand, the sex ratio would be little altered if we were able to choose the sex of our children (Hartley and Pietraczyk, 1979).

In humans, as with a lot of other animals, men and women do not reproduce evenly. Whereas most females mate successfully, the males enjoy a highly varied success rate. The strongest, healthiest, and most socially powerful men have access to a disproportionately large number of women and (disregarding for the moment the effect of modern contraceptives) pass their genes on to a greater proportion of the following generation. This, of course, is one of the prime bases of sexual selection – a major way in which the species changes as a whole towards greater adaptability, and a procedure by which the males become progressively differentiated from the females. (It was Darwin who pointed out that the males in most species are the more modified sex. They are the ones who change the most during development and appear to be more specialized in body and function.)

If the superior males are the ones who are most successful at reproduction, what happens to the others? What do you do if you are a male who is missing out? The most extreme solution is suicide, and while I do not argue that this is an adequate explanation of most male suicides, one could consider its evolutionary significance with respect to competition between males. Other males are, of course, killed in combat, which can probably be traced eventually to competition for territory, if not females. But if the man continues to live he must find some other means of dealing with his sex drive. One way is to turn it off via the brain-hormone connection – and one wonders to what extent this is involved in the etiology of depression. Certainly, depressed patients show a fairly striking diminution of sex drive, though of course illness of any kind usually results in a loss of sexual interest. This makes good evolutionary sense too, since it is better that the healthy members of the species should be involved in breeding than those that are susceptible to sickness. Other adaptions include masturbation, homosexuality, and the other sexual deviations which involve little likelihood of procreation. Practices such as fetishism, bestiality, exhibitionism and voyeurism, which are so deplored by our society, may actually be serving the important sociobiological role of hiving off superfluous male libido. This point will be dealt with in greater detail in Chapter 7.

Since it is more important for a man to be competitive than a woman, we might expect natural selection to have arranged things so that male children would be born to women more often when they are in a prime, healthy condition, whereas girl babies would be produced when their condition declines. Such an effect has been documented in a variety of species including humans (Wilson, 1975), although the probable mechanism is that male babies are more susceptible to disease *in utero* and suffer a higher foetal mortality than female babies.

Changing sex after birth

Some species have a very clever way of dealing with the problem of imbalances in the sex ratio. The individuals are able to change their sex for the sake of social or reproductive convenience. One species of fish found in the coral reefs of the tropical Pacific forms social groups consisting of one male and a harem of females occupying a unit of territorial water. The male supresses any inclination of the females to undergo a sex change by aggressively dominating them, until he dies, at which point the dominant female in the group promptly turns into a male and takes over the harem (Robertson, 1972). While it may seem a considerable leap from fish to us humans, we also associate masculinity with social ascendancy to the extent that some males who have difficulty competing with members of their fellow sex find it a great deal more comfortable to adopt the female role and identity, sometimes to the extent of having a sex change operation. It seems clear that in the case of most transsexual men it is not so much a desire to experience sex from the female point of view that motivates them as an absence of the libido that would affirm their manhood. They feel a great deal more relaxed when assuming the female role or affter being reassigned to womanhood (Gosselin and Wilson, 1980).

Adult women who wish to become men are relatively rare. However, in a particular remote mountain area of the Dominican Republic, quite a few girls (about 37 in the past 50 years) have been observed to turn spontaneously into men around the age of puberty (Imperato-McGinley *el al*, 1974). At birth they appear as normal females but around the age of twelve they begin to develop a penis, testicles, and all the secondary sex characteristics of a man. They are apparently able to adapt to their new sex identity perfectly well and enjoy a satisfactory sex life, although they are infertile. Family tree studies of this group have shown they are descended from a common ancestry, so presumably they are the victims of a mutation that is unlikely to have much of a future because it has no evolutionary value.

Supermen and superwomen

The sex of a person is basically determined by the shape of their twenty-third pair of chromosomes. In the normal woman, these two chromosomes, when highly magnified, look much the same (like a pair of X's) and this is described as an XX pattern. In men, one of the pair is much smaller than the other, and the pattern is known as XY. While most people show one or the other of these two patterns, a wide variety of sex chromosome anomalies have been discovered from microscopic examination of people's body cells, some people being found to have more than the normal two sex chromosomes. Generally speaking, the individual will appear as female so long as she has only X chromosomes in her set, and will appear as male if there is at least one Y chromosome in the set. Now a certain proportion of men (about one in a thousand) have been discovered to have two or more Y chromosomes in their body cells (usually an XYY pattern) and have therefore been dubbed 'supermen'. Although they are not necessarily identifiable as different from normal (XY) men, there is apparently an increased likelihood that they will be exceptionally tall and aggressive (with the result that the prison population contains a slightly elevated proportion of such men). Note that both these characteristics – height and aggression – are typically male as opposed to female traits.

What characteristics might we expect to discover in 'superwomen', that is, women with chromosome combinations of XXX, XXXX, etc.? It seems that with each additional X chromosome, the woman is more likely to be fat and mentally retarded (Dewhurst and Gordon, 1969). Unfortunately, there is no information concerning personality characteristics such as nurturance.

Whilst it might gratify male vanity to suppose that fatness and mental retardation are exaggerations of typical female characteristics, just as height and aggression are typically male, this would be unfair. A pure X pattern really corresponds to the unmodified human being; it is Y chromosomes that make for changes from the standard form. Therefore, we might argue that additional Y chromosomes could enhance the extent of the modification from the standard human form, whereas no amount of adding Xs would increase femininity *per se* (however much ideal bodily functioning could be perturbed). In other words, what I am suggesting (even though it spoils a good story) is that chromosome anomalies tell us something about the essential nature of masculinity but do not necessarily tell anything about the nature of femininity. Women without sex chromosome anomalies are, in fact, less likely to be mentally retarded than men, so retardation can hardly be called a typically feminine trait.

Genetic variation in masculinity-femininity

One of the best reasons for supposing that the average differences between men and women are biological is the discovery that variations in masculinity versus femininity of personality measured within each sex are strongly influenced by genetic factors. That is, when a questionnaire designed to assess traits like promiscuity, aggressiveness, independence, and so on, that are traditionally supposed to separate men from women is given to a sample of same-sex identical twins and a comparable sample of same-sex fraternal twins, the identical twins produce scores that are much more alike (Nichols, 1978). Recent estimates suggest that as much as one half of variation in masculinity versus femininity is due to genetic causes (Eysenck, 1976). This figure again applies to variations occurring *within* sexes, not between them.

Such work highlights a point of view that is gaining increasing currency – there are not just two sexes, but a range of people who are male or female to a greater or lesser extent both anatomically and mentally. These differences are laid down by the balance of hormones circulating before birth and it is fortunate that for the most part they push people into one camp or the other sufficiently for society (doctors, parents, priests, etc.) to make categorical decisions as to whether each child is male or female. This categorization is cruel to some individuals who fall towards the middle and are thereafter expected to conform to a stereotype that does not suit them. In other cases it is just plain wrong. Nevertheless, it remains a reasonably convenient fiction.

The role of hormones

It is now widely known that the differences between men

and women in body and behaviour are mediated to a large extent by hormones. The sex hormones are a group of closely related chemicals that each have an individual shape which allows them to act like door keys, fitting into their own receptor sites in the lower part of the brain which are shaped to receive them. These receptor 'locks' can therefore only be operated by their own particular hormone; if that hormone is not available in the blood-stream or if the site has been blocked by some impostor, then that part of the brain will not be activated. The effects of hormones, therefore, depend both upon their concentration in the blood and the sensitivity of the receptors.

The most important of these hormones are the *oestrogens* which cause the female to be receptive and physically desirable to the male, *progesterones*, which prepare for and support pregnancy, and the *androgens* (particularly *testosterone*) which make us sexually assertive. The latter are, of course, much more prevalent in males (being produced in large quantities by the testes) but the small amount that is present in females (produced mainly by the adrenal cortex) is probably responsible for the assertiveness that women do exhibit. Part of the evidence for the various functions of these hormones comes from studies of the effects of administering to people of one sex hormones that are more normally associated with the other. Thus, if women are given injections of testosterone, they are likely to show higher levels of libido and aggressiveness (two traditionally male characteristics). Because of this, testosterone has been used to treat women who complain of insufficient sexual responsiveness. However, there is a danger that if too much is given the woman may begin to develop masculine physical characteristics like hair on the chin and an enlarged clitoris. If oestrogens are given to men, the reverse takes place; their normally high libido and aggressiveness are reduced. For this reason oestrogen has been used to treat sex criminals, although of course, there is the problem that the man may grow breasts and put on fat around the hips. The British penal authorities have recently been successfully sued by a prisoner who suffered these effects.

The fact that oestrogen increases female receptivity while at the same time having an inhibitory effect upon male-style sexual assertiveness may help to resolve a great deal of confusion that has arisen concerning variations in the 'sexiness' of women at different times of their cycle. Oestrogen secretion peaks twice during the female cycle, once shortly before ovulation and then again in the middle of the post-ovulatory phase. Progesterone secretion begins around ovulation and increases to a post-ovulatory peak. If pregnancy has not occurred it then declines to zero. Testosterone is produced in such small quantities in women that it is difficult to monitor, but it is believed that it is produced fairly evenly throughout the cycle, with possibly a slight peaking around the time of ovulation.

Studies of women's sexual activity throughout the cycle suggest two peaks of libido, one at ovulation and another around the time of menstruation. How could the patterns of hormone secretion account for these? The enhanced sexual activity at ovulation could be due to the slight peak in testosterone at that time; on the other hand, the high levels of oestrogen (even though falling by that stage) would probably offset such an effect. More likely, the increased frequency of intercourse that has been observed in the female mid-cycle is due to increased receptivity and attractiveness of the female to the male partners rather than any increase in female assertiveness. The oestrogen peak in the post-ovulation phase would not have the same effect on receptiveness because it is smothered by very high levels of progesterone, the hormone which maintains pregnancy and which appears to work against sexual activity to some extent. Around menstruation, however, both oestrogen and progesterone cease production, perhaps leaving testosterone free to evoke some male-like sex drive. Female sexiness then, is probably two-dimensional, with the two cycle peaks that have often been observed differing in quality. The mid-cycle peak could be described as reflecting hyper-receptiveness, while the menstrual peak may be better thought of as freedom from the usual female inhibitions and more of an approximation to the typically male 'hunting' instinct.

It should not be thought that the relationship between testosterone and libido is a simple one. Levels of testosterone in the bloodstream vary enormously from time to time in the same individual, and while they sometimes seem to promote sexual activity, they may also change as a consequence of sexual activity. Normally, there is an increase in production of testosterone in a man after a satisfactory sexual encounter, perhaps in order to 'recharge the batteries' for further action. There is a similar reciprocal relationship between testosterone and social power. High levels of testosterone may promote aggression and dominance, but the outcome of struggles for social ascendancy also affects testosterone level (Rose, Bernstein and Gordon, 1975). If Borg and Connors were to have their testosterone measured before a Wimbledon final, and then again shortly afterwards, it is fairly certain that the victor would show an increase over his pre-match level while the loser would show an equivalent decline. In this way, the man who is successful in competition with others is prepared to capitalize upon it, the usual reward being female sexual favour. A possible evolutionary function of this might be to mediate between male strength, skill and social power on the one hand, and access to females, potency and fertility on the other, with the result that superior males achieve greater reproductive success.

This might also explain the indulgence of conquering armies in horrific acts of rape as well as pillage. Perhaps the elating experience of conquest in battle leads to a surge in testosterone which is manifested in brutish sexual assaults on the women of the defeated populace. This, of course, does not justify such behaviour in any way. A distinction has to be made between physical and moral superiority.

Prenatal setting of sexuality

The sex hormones begin to affect our behaviour before birth. As any parent with a mixed-gender family will attest, little boys are unmistakably different in temperament from little girls virtually from the moment of birth. This difference comes about because the male embryo has testes which produce testosterone, and this hormone affects the development of all the bodily structures of the baby including the brain. Striking direct evidence for this effect has been provided by June Reinisch of Rutgers University, New Jersey. In an article published in *Nature* (1977) she described her investigations of the personality of boys and girls whose mothers had been treated with steroid hormones for one reason or another during pregnancy (usually to maintain a pregnancy that was at risk). Children whose mothers had been given progestines (synthetic progesterones which simulate androgens in many ways) during pregnancy were found to be more independent, self-assured and less dependent on the group than those whose mothers had been treated with oestrogens. The same result applied whenn the 'treated' children were compared with untreated siblings.

Another impressive demonstration of the influence of prenatal hormones comes from the work of the German physician W. Schlegel (1975). The sex hormones determine that male and female skeletons are constructed differently, the most obvious difference being a broader outlet to the female pelvis (for the baby to pass through during childbirth). Now x-rays reveal considerable variation in the width of the pelvic base *within* each sex, some women having a male-type pelvis and some men having a female-type pelvis. Schlegel's contribution was to show that social and sexual behaviour was correlated with the maleness or femaleness of the pelvic shape within each sex. Men and women with pelvises that were atypical of their own sex were more likely to show the traditional behaviour and sexual propensities of the opposite sex. They were also more likely to be homosexually oriented and to have suffered a marital breakdown. The easiest interpretation of these findings is to say that hormones circulating before birth simultaneously lay down skeletal features and brain mechanisms relating to sex-role behaviour so as to be compatible. Although most people are easily classified as either male or female following this process, there are great variations along a continuum of masculinity-femininity within each gender.

Having previously said that current levels of testosterone may affect sex drive, it is important to add that to a considerable extent an individual's characteristic level of libido is constitutionally fixed. Just as Eysenck (1976) demonstrated with twin studies that masculinity-femininity of personality and attitudes are partly determined by heredity, so too he was able to conclude that a good part of the variation in libido within each sex was genetically determined. Taking, for example, a single index of libido, the age at which a girl first engaged in intercourse, he found that identical twins have their first experience at much closer ages than non-identical twins.

Experiments with subhuman species are also illuminating. If male rats are classified with respect to their characteristic frequency of copulation and then castrated, the sexual activity of both high and low libido groups is reduced to near-zero. Then, if they are given replacement injections of testosterone in massive doses, much beyond that which occurs naturally, the various rats return only to their previous level of sexual activity. Those that were highly active before castration are reactivated to the same high level, while those who were fairly uninterested revert to their same low frequency of copulation (Bermant and Davidson, 1974). Apparently, male sex drive is to a large extent set by brain mechanisms which only require a certain threshold level of testosterone for full operation. This explains why the sex drive of normal men is not increased by injections of testosterone – they are already above threshold in this respect. However, the sex drive of women, or men who are deficient in testosterone for some reason, is enhanced by additional testosterone.

Male and female superiorities

Darwin noted two ways in which sexual selection could occur. One is competition between members of the same sex which would give certain individuals greater access to the opposite sex; the other is the preferences of one sex for certain traits in members of the other. Generally speaking, it is the males who compete among themselves for dominance and control of territory, and consequently the privilege of mating with the females. For example, if a man is much stronger than his rivals he may be able to scare them away from any girls that he fancies and thus have intercourse (and children) with those girls. Such a pattern would lead to a progressive increase in muscular strength, possibly in the human race as a whole, but especially in men. The result is that in most higher animals the male is bigger and stronger than the female. This is particularly true of species in which the dominant males have more than their fair share of the females – which is to some extent the case with human beings. If a species is monogamous in its mating behaviour, so that the males do not compete for the females, then the size difference is often small or non-existent.

With modern man, of course, brute strength is of less importance than intellectual cunning and resourcefulness when it comes to securing more than average access to women. One might therefore expect that, particularly in humans, the sexes would diverge with respect to ingenuity. Darwin himself believed that men were more persevering and inventive than women, and for this very reason. Men have to use their brain not just to stay alive but also to compete for women, and this Darwin held to be the basis of the predominance of male genius in the world over that of females.

However, there is a reason why this argument does not follow quite as well as it may seem to, and to be fair to him, Darwin himself pointed this out. Generally speaking, it seems that the two sexes of species will tend to evolve similarly unless there is a good reason why they should not. Since intelligence is unlikely to be a disadvantage to members of either sex, we would expect to see it develop in both sexes even if it was the males who needed it most. On the other hand, the physical strength, aggressiveness and courage of men have evolved at the expense of characteristics which in women are more necessary, such as endurance, nurturance and a care for security. We would therefore not expect that, in this emotional field, male attributes would transfer to women in the same way.

The distribution of intelligence between the sexes can be checked empirically by IQ tests. Huge numbers of such tests have been administered, of course, and they do not register any overall difference. The advantage of men on the spatial and mathematical problems is offset by the advantage shown by women on verbal and rote memory tasks. But there is a greater spread of scores for men which suggests that higher proportions of them could be classified as having IQs at the level of genius and as being mentally retarded. This ties in with the general tendency of men to show more extremes and abnormalities of every kind than women, and is probably due to the fact that males lack a second X chromosome which could offset anomalies on the first. Hence they are more susceptible to genetic disorders such as colour blindness and haemophilia which are located all or in part on the X chromosome.

Exceptionally high intelligence and specific talents like artistic, musical and mathematical ability could be construed as freak phenomena that are equivalent to these other abnormalities. Indeed, autism and childhood schizophrenia seem to have links with maleness, genius and mental retardation, all three. Alternatively, it may turn out that Darwin's view is correct and that male genius is not so much an example of genetic abnormality, manifested more frequently in men than women, but rather has been built into our species as a result of sexual selection. In other words, the preponderance of male genius may be viewed as analogous either to haemophilia or to the peacock's tail display. Both of these models appear more plausible than that of 'social role learning' which is currently so fashionable.

Physical beauty

The other important principle of sexual selection that would be expected to give rise to sex differences according to Darwin is that of variations in attractiveness to the opposite sex. Apart from the sheer competition for mating with large numbers of the opposite sex, there are the preferences that we feel and exercise for different representatives of the opposite sex. Thus if most women preferred tall men we would expect men to become progressively taller than women, and if men preferred women with large breasts we would expect the breasts of women to evolve progressively larger than those of men. Presumably there are limits to this process set by the previously mentioned tendency for selected attributes to be shared across both sexes (especially in mammals) and by the limits set on our biological practicability. Men who are greatly taller than average become inefficient machines because their limbs are gangling and weak and their heart has difficulty pumping blood to the high altitudes of their brain. Women whose breasts are over-large have difficulty with support and mobility; and anyway, if the breasts droop they suggest old age and cease to be seen as attractive.

Darwin confessed himself somewhat puzzled by the tremendous cultural variation that he observed in the criteria of beauty among various societies. Whereas we Europeans value a good set of white teeth, there are cultures in Africa and Indonesia where they stain their teeth black or even knock them out so as to appear less 'brute-like'. And whereas scars are thought villainous and unattractive in our culture, there are countries in which nobody is thought beautiful until their face has been gashed or mutilated in some way. Such differences in the appeal of various attributes could be partly at the root of racial differentiation, for whenever a quality is preferred it will become emphasized by the process of sexual selection. One rule that Darwin did observe was that where a racial group was distinguished by some unusual attribute, this characteristic was usually rated by that group as attractive. Hottentot women have bottoms that jut out conspicuously, and according to early explorers the women in whom this characteristic was most outstanding were regarded by their menfolk as the greatest beauties. Similarly, those tribes in the South Pacific who are generally bearded value a profuse and bushy beard while others who are relatively beardless go to great lengths to remove what little facial hair they produce. Darwin concluded that 'man admires and often tries to exaggerate whatever characters nature may have given him' (1871). This presumably leads to progressive divergence among the races, and could perhaps eventually lead to the development of separate species.

The other general rule that Darwin deduced from all the anthropological evidence he could assemble was that humans are all greatly concerned about their looks, and that women are usually more conscious of the way they dress and decorate themselves than men. This would follow if it is true that selection among males is mainly determined by successful aggression or cunning; the dominant men would then make their choices among the women on the basis of physical attractiveness. This, no doubt, is the reason for the well established fact that the looks of a woman count for more than those of a man in the choice of partners (Wilson and Nias, 1976).

One interesting piece of evidence relating to the physical attractiveness of women that was first pointed out by Darwin and later taken up by Desmond Morris (1971) is that of the loss of hair in humans relative to other apes. Clearly, this is a late development in evolutionary history and since it is in some respects detrimental to our chances of survival (especially through cold winters) it is hard to think of any reason for it other than its cosmetic function. It seems to have developed first in women, presumably because the exposure of flesh was attractive to men, and although this characteristic has been transmitted to men as well to some extent, hairlessness is much more obviously characteristic of women. This would suggest that the looks of women were leading the evolution of our appearance for the most part, which ties in with the fact that in monkeys it is a female visual display – the presentation of her heat-swollen backside, to the male – that is the primary stimulus to sexual activity. Of course this invitation may in turn be partly prompted by visual attributes of the male, but to a large extent it seems to be determined by the internal hormonal state of the female as linked to her cycle.

The looks of men also seem to have changed as a result of sexual selection. Darwin cites evidence that the male beard, which is also seen in some other primates, is not just a remnant of our earlier hairiness, but an ornament that has evolved positively in order to impress or attract women. The fact that not all women in our culture like men with beards is no case against such a view because of the complex social meanings that attach to beards today (political radicalism, Bohemianism, etc.) A dislike of beards may be attributed to these causes rather than to an instinctual disgust at hairy male chins. In addition, we need to take into account the previously mentioned fact that the women from racial groups that are relatively beardless are likely to find beards less attractive than a smooth chin.

Because the looks of women seem to be more important than those of men it should not be thought that women play no part in selecting mates. They are at least equally important agents of selection, and perhaps even more influential than men. However, the criteria on which they base their selection are different. Women choose men on the basis of physical strength, height, courage, skill and intelligence more than on visual features.

Civilization and competition

One might argue that all this is irrelevant since in modern civilized society we have adopted a monogamous system of one man, one wife, and that all individuals have equal chances to transmit their genes through families. Well for one thing, this is not entirely true. Men still have mistresses and get divorced in European society, and there is little doubt that such activities are more common with respect to men who are high in the social scale. Men who are rich, powerful, successful and talented find it a great deal easier to 'pull women', if that is their interest, and they can also afford more easily to get divorced and remarried and to start second families after they have become widowed.

There is evidence from blood tests and more recently from the new technology of tissue-typing that quite a high proportion of children (perhaps as much as one third according to one survey in suburban London) are sired by a man other than their presumed father. This being the case, one wonders about the characteristics of the other men by whom our suburban housewives are getting themselves pregnant, either accidentally or wittingly. Presumably, they are successful and dominant men on average, at least more so than either the rightful husband or the milkman. The only counter factor in all of this is that the more intelligent men are probably also more considerate and skilled in the use of contraceptives, which might have been offsetting the effect to some extent in recent decades.

But even if our society was perfectly monogamous today, this would not alter the manner in which we have evolved and the biological instincts that we have inherited as a result. It would take thousands of years at the very least before our current social behaviour, or the effects of contraception, could be expected to manifest themselves in changes in human personality and motivation. In the meantime, it remains constructive to regard ourselves as slightly modified apes with behavioural dispositions appropriate to our development on the African savannas or even earlier. These dispositions are different for men and for women in various respects which have been outlined in this chapter and which will be elaborated in those that follow.

3
The Double Standard

One of the central complaints of the current feminist movement is that society espouses different standards of sexual morality for men and women. Certainly, the fact of this double standard is unquestionable. Surveys in both Britain and America (for example, Hunt, 1974) reveal that premarital sex, adultery and promiscuity are regarded as greater sins in women than in men. It is also clear that this attitude is not imposed by men on women but is shared equally by men and women. Somehow, lustfulness and the seeking of sexual variety are relatively tolerated in men because they are seen to be more powerful urges, so integral to male nature that men can hardly be held responsible for them. Women, on the other hand, are seen to be naturally virtuous as regards matters of sexual propriety, and their transgressions are therefore treated as less excusable.

There is little doubt that these differences in sexual inclination between men and women do exist. An enormous amount of research from Kinsey onwards has confirmed the reputation of men for seeking multiple partners and impersonal sexual thrills, while women look to exclusive, 'meaningful' relationships with particular men. While men typically try to have sex with a girl first, and only afterwards consider whether she is interesting and intelligent, women prefer to get to know the man well and build up a relationship of trust and understanding before engaging in physical sex. Much of human courtship behaviour can be understood as an interaction between these opposing strategies of the two genders.

As an example of the kind of research that has established these differences, Eysenck (1976) administered a lengthy questionnaire concerning sexual attitudes and inclinations to several hundred men and women in Britain. Considerable gender differences were revealed with respect to issues such as extramarital sex, mate-swapping, orgies and pornography, the men emerging as more libidinous and adventure-seeking. Similarly, in a study of sex differences in fantasy patterns, Wilson (1978) found that men were much more prone to fantasize group activities and sex with anonymous partners, while women dreamt of romantic, intimate liaisons with particular identifiable men.

Extramarital desires

Studies of the occurrence of extramarital sex also reveal male/female differences. The Kinsey research, and studies in a variety of Western countries since then, have shown that something like one half of married men and a quarter of married women commit adultery at some time in their life, although the most recent studies have shown some upward movement in these figures especially for women. Since the gap between men and women appears to be closing, these figures might seem to suggest that there is not a lot of difference in their inclinations. Such an interpretation would be quite mistaken however. Since each act of adultery logically entails at least one man and one woman, there is considerable pull towards a levelling of the figures. In the past, the difference in extramarital experience between men and women must have been partly due to prostitutes and single girls having sex with married men. With a decline in the oldest profession and a lowering of the marriage age it is not surprising that the sex difference in rates of extramarital activity is diminishing.

The real difference between men and women in extramarital sexuality is not so much in the behaviour as in the desire and the motive. When surveys are worded so that people are asked if they would *like* to engage in extramarital sex, the male/female difference is greatly enhanced. A very high proportion of men, but very few women, are attracted to the idea as a theoretical proposition. It appears that men are prevented from having extramarital sex mainly by lack of opportunity, whereas women are more often inhibited by their own attitudes, morals and preferences. After all, if a woman does feel like playing she generally has little difficulty in gaining male co-operation: she has only to step into the street and call 'Take me now'. A man adopting the same approach would be unlikely to attract anybody except the police.

The other difference between men and women with respect to adultery concerns the reason for engaging in it. Symons (1979) has assembled a considerable amount of evidence to support the idea that whereas men frequently engage in extramarital sex for the sake of novelty and excitement, these motives are seldom sufficient for women. When women are questioned about why they had extramarital sex, it usually turns out that their relationship with their husband was in some way unsatisfactory at the time or that they saw the other man as being in some sense superior to their husband. The simple fact that the other man was different is hardly ever offered as an excuse. Another fairly common reason given by women for having committed adultery is that they hoped to create some effect on their husband, for example, getting revenge on him for similar behaviour, or stopping him from taking them for granted. Again, men very seldom engage in sex in order to make an impression on their wives; usually they are at considerable pains to prevent them from finding out.

Despite the fact that men are more inclined to indulge in extramarital activity than women, and actually do so at some time or another, sexual jealousy in the husband is more often a source of marital difficulty than the suspicions of the wife. Whereas about half of all divorced men cite sexual infidelity on the part of their wife as a major reason for the breakdown, only about a quarter of women do so. And even after the grounds for divorce in Britain had been made equivalent for men and women, the 1950 Matrimonial Causes Act left a divorcing husband, but not the wife, the right to claim damages from a corespondent. Scottish husbands retained this right until 1977 (Daly and Wilson, 1979).

Such differences reveal a striking 'double standard' of sexual morality which survives today in Western society and which is subscribed to equally by men and women. Historical and anthropological comparisons reveal that similar attitudes have prevailed in virtually every time and place. Among the ancient Greeks, for example, it was usual for a husband to have sex with prostitutes, foreign women and slaves, while the wife was expected to be a virgin before marriage and faithful to her husband thereafter (Bardis, 1979). Men were permitted to have concubines as long as they could afford them and jealousy on the part of the legitimate wife was to some extent averted because her status (and that of her children) was higher than that of the concubine.

Polygyny, the marital system by which a man may have several wives, is by far the most prevalent arrangement among human societies. George Murdock's *Ethnographic*

Atlas presents a categorization of marital systems in 849 societies; of these, 709 are polygynous and only four are polyandrous (women being permitted more than one husband). Brides are widely treated as acquisitions; they may be bought and sold, inherited and loaned out in order to repay debts. This is not simply exploitation, it is also a reflection of supply and demand. In all societies prostitution is almost exclusively a female occupation. Male prostitutes, as well as male strippers, would soon be out of business without their homosexual male custom. Sex is widely regarded as a service that is rendered by females to males, while marriage is seen as a means to ensure the economic security of the woman and her offspring.

'Wanton women'

Although there are, on average, differences between men and women as regards interest in sexual adventure and multiple matings, there is also some overlap between them. A certain proportion of women do enjoy sexual diversion and some men are disapproving. It follows that any rigid expectation that a person will behave in a particular way simply because of his or her gender is bound to be unfair on those individuals who do not naturally conform to their stereotype. To this extent at least, the feminists have a legitimate grievance. It is unfair to assume that because a person is male or female they must behave in a certain prescribed way, and if they do not they are morally reprehensible.

Many societies have laws against adultery under which women are much more severely penalized than men. Some readers may recall the horrific scene in the film *Zorba the Greek* in which a young Cretan woman is stoned to death for sleeping with a visiting Englishman even though she has been widowed for some time. In the famous myth of ancient Rome, Lucretia found it necessary to kill herself to defend her husband's honour because she had been raped by an Etruscan Prince. We still hear today of women in Islamic countries such as Iran being executed for adultery and similar treatment is occasionally meted out in Oriental countries. In all of these cultural contexts, the same misbehaviour on the part of a husband would be treated relatively lightly. Such discrimination is clearly quite unfair, even though the social and biological background to it be understood.

Psychiatrists in the past have been guilty of much the same kind of injustice, labelling as abnormal any person who was too far removed from the gender stereotype. They were particularly harsh on women who displayed active and adventurous sexuality of the kind that would probably be applauded as a sign of social success in a man. The following quotation from the work of the influential Victorian psychiatrist, R. Von Krafft-Ebing will serve to illustrate the prevailing views of such women among the medical profession:

> CASE 190. *Chronic nymphomania*. Mrs E., age 47. An uncle on father's side insane. Father suffered from self-conceit and was given to sexual excess. A brother of the patient died from acute cerebral inflammation. Always nervous, eccentric, erotic, began coitus at age of ten. Married at nineteen. Although her husband was virile, she maintained a number of male friends. Fully conscious of the abominable nature of her conduct, she was powerless in restraining her insatiable appetite. She kept up appearances, however (*Psychopathia Sexualis*, 1886).

Within the social context of the day, this poor woman might just as well have been afflicted with plague or leprosy. Today, it is doubtful whether she would be distinguished as neurotic at all; she might even be the envy of her female friends.

The worthy doctor goes on to describe the seriousness

of her condition in a manner that is hard to distinguish from a sermon:

> Chronic conditions of nymphomania are apt to weaken public morality and lead to offences against decency. Woe unto the man who falls into the meshes of such an insatiable nymphomaniac, whose sexual appetite is never appeased. Heavy neurasthenia and impotence are the inevitable consequences. These unfortunate women disseminate the spirit of lewdness, demoralize their surroundings, become a danger to boys, and are liable to corrupt girls also, for there are homosexual nymphomaniacs as well.

The similarity between this analysis and that of the medieval witch is quite striking. Witches too, were reputed to drain sexual energy out of men.

Elsewhere Krafft-Ebing makes it clear that he also regards adulterous urges towards a particular man to be pathological in a woman.

> A particular species of excessive sexual urge may be found in females in whom a most impulsive desire for sexual intercourse with certain men imperatively demands gratification .... This pathological want becomes so powerful that all consideration of shame, conventionality and womanly honour simply disappear, and it reveals itself in the most shameless manner even to the husband, whilst the normal woman, endowed with full moral consciousness, knows how to conceal the terrible secret (1886).

Although Victorian psychiatrists also recognised pathological conditions of excessive sex drive in men, it is quite clear that their criteria for assigning such diagnoses were very different. Because hungry and promiscuous sexuality was observed to be unusual in women (no doubt particularly so in their day) when it did occur it was much more likely to be attributed to pathological causes than the equivalent behaviour in men. In fact, in men, a lack of interest in sex was much more likely to be identified as pathological.

Krafft-Ebing's views on the essential differences between men and women, which no doubt are basic to his views on the abnormality of 'nymphomania', are interesting in that they express a conventional wisdom that is by no means unique to Victorian culture.

> Man has beyond doubt the stronger sexual appetite of the two. From the period of pubescence he is instinctively drawn towards woman. His love is sensual, and his choice is strongly prejudiced in favour of physical attractions. A mighty impulse of nature makes him aggressive and impetuous in his courtship. . . . Woman, however, if physically and mentally normal, and properly educated, has but little sexual desire. If it were otherwise, marriage and family would be empty words. As yet the man who avoids women, and the woman who seeks men are sheer anomalies (1886).

Social learning theory

Krafft-Ebing's statements seem ludicrously oversimplified to the modern reader. We are no longer so inflexible in our thinking about how men and women ought to think and feel sexually. Although it might not have disappeared completely, the double standard in its various manifestations has beaten an unmistakable retreat since the turn of the century.

This observation of variability over time and culture has led some social scientists to propose that all such conceptions of male and female sexuality are socially conditioned. In recent years it has been increasingly common to encounter the belief that the only biological difference between men and women which has any bearing on their sexual behaviour is the difference in size. Men, being so much bigger than women, have gained a monopoly on social and economic power and bullied them into subjugation and sexual slavery. The double standard has arisen, according to this view, because men have imposed their own selfish will upon women. If women were stronger than men the position would be reversed; most societies would be polyandrous rather than polygynous and men would be more severely punished for promiscuity and infidelity. No doubt they would also have to do the housework, cooking and child-minding, while women pursued exciting careers and casual sex with men.

This argument, which has been taken up by many feminists, implies that the evolutionary accident which gave men the advantage in power has resulted in a world-wide male conspiracy by which little boys and little girls are programmed from the cradle to adopt behaviour conforming to their gender stereotype. Boys are taught to fight, assume independence and responsibility, and to treat girls as objects of sexual pleasure, while girls are taught to be submissive and play with dolls in preparation for the mother role. According to this view, the double standard would cease to exist if boys and girls were raised in exactly the same way and taught the same 'healthy' attitudes towards love and sex. Some moves have been made towards this objective with the screening of fairy tales and school books for 'sexist' content and with media pressure on parents and teachers to treat boys and girls as identical creatures.

Unfortunately, social learning theory does not provide an adequate account of the full extent of mental and behavioural differences between men and women. While it is no doubt possible to change social attitudes to some extent with appropriate persuasive techniques, there are reasons for thinking that the double standard in sexual morality will not be so easily disposed of. Even if temporarily overridden it is likely to re-emerge as soon as social pressures are relaxed. We have already seen in Chapter 2 that the traditional differences in personality and predisposition between males and females are partly determined by genetic factors operating through the sex hormones. The difference in size between men and women is not the cause of the respective sexual tendencies; it is just as much a consequence of them. As I hope to demonstrate in the rest of this chapter, the physical and mental characteristics of the two sexes have evolved in parallel, in order to serve their respective strategies for reproductive success.

Parental investment theory

The key to understanding the so-called double standard, and for that matter most of the differences between men and women in the area of love and sex, is to be found in the concept of 'parental investment' that has been developed most notably by the Harvard biologist, Robert Trivers (1972). This theory starts from the assumption, introduced in the previous chapter and now widely agreed upon by biologists, that the behaviour of an animal, like its body, has evolved in order to promote the survival of the genes carried in each of its cells. Thus the individual has instincts concerned not just with his or her own personal survival, but with the welfare of all the relatives and descendents with whom some of his genes are held in common. Obviously, the sexual instinct serves this purpose, and so does the maternal instinct.

Trivers' central idea is that the two sexes invest differently in their offspring. In mammals, for example, the male invests only a small amount of time in courtship and copulation, whereas the female has to devote a great deal of additional time and energy to pregnancy, childbirth and lactation. As a result, the number of offspring that the female is able to produce in her lifetime is very much less than that which the male is, theoretically at least, capable of producing. The male can greatly increase his reproductive success by mating with a large number of different females, whereas the female has very little to gain by multiple matings. Her main interest is in guaranteeing the survival of the child that she does bear.

The distinction between men and women in terms of child-care encumbrances has an analogy, if not actual roots, in the comparison of sperm and egg. In most animals the egg is large, static and nurturant while the sperm is small, mobile and penetrating. There is a sense in which the sperm is the hunter and the egg is the target. Once fertilized, the egg provides care, protection and sustenance to the developing embryo, while the job of the sperm is completed. The maternal care provided by the mother after the birth may be conceptualized as an extension of the biological protection that the mother's womb provides to the embryo, while the hunting of the male could be seen (perhaps rather more metaphorically) as an extension of the behaviour of his sperm within the female reproductive system.

Because eggs are larger and more complex than sperm they are very much at a premium compared with sperm, which are tiny, mass-produced, and expendable. The human female produces only a few hundred eggs in the course of her lifetime, released one or two at a time and spaced at monthly intervals. The male produces literally millions of potentially fertile sperm with each ejaculation, and this may occur several times a day. It follows that a woman has much more reason to be concerned about the outcome of her eggs than a man does about the fate of his sperm. The woman has very few chances to reproduce her genes, but the instinctual logic of a man may understandably take the position that 'there is always more where that came from'.

Another important difference between eggs and sperm is that eggs are present in the woman's body from birth and are therefore susceptible to environmental damage of one sort or another, such as radiation or chemical interference. This means that as a woman gets older there is an increased chance that her offspring will be deformed in some way. While mutations are occasionally beneficial to the species as a whole, the vast majority of them are detrimental, so any individual woman is generally better off having her babies when young. A man's sperm are little altered by his age since they are manufactured fresh on each occasion. As he gets older he becomes slightly less virile and less fertile because his ejaculate is less concentrated and his orgasms more widely spaced, but he remains theoretically capable of producing healthy children until his dying day. To take the best known example of genetic damage to human babies, Down's syndrome (Mongolism) is much more likely to occur if the mother is approaching menopause, but the age of the father is fairly irrelevant.

On the basis of all these considerations we can arrive at some major predictions about the way males and females could be expected to behave.

1 Polygamy: Males should tend towards multiple mateships, while females will prefer to cultivate a particular mate whom they will try to persuade to assist them in the process of rearing offspring.

2 Competition: There will be more competition among males for access to females than vice versa, with the result that traits favouring success in competition, such as strength and skill in developing weapons, will be more highly developed in males.

3 Coyness: Females will be concerned to investigate the quality of the male they have sex with, requiring demonstrations of mental and physical fitness, and perhaps also an indication of long-term devotion, before submitting to copulation. This is presumably the basis of the elaborate courtship rituals that are seen in many species, with the male normally 'making the play' and the female showing reserve and caution.

4 *Age differential*: Males will be attracted to younger females who are in prime breeding condition whereas females will be relatively unconcerned about the age of their mate provided there is some evidence of his past fitness. This is probably a consequence of the differing effect of age on the viability of eggs and sperm, and will be discussed more fully in the next chapter.

The polygamous male

The male preference for polygamy is but one reflection of their desire for variety in partners. Provided that not too much time and effort is required for the seduction, it is to the reproductive advantage of men to have casual sex with as many partners as possible. From the point of view of the 'selfish gene', there is always a chance that a pregnancy could result, and even if the father has moved on geographically or emotionally the woman will still be motivated to raise the offspring alone or with the help of any new mate that she can acquire. The reader should not be confused by the fact that the conscious motives of the parties may appear to be in direct opposition to the sexual instincts involved. In the absence of religious strictures to the contrary, both men and women will usually opt to use contraception in connection with casual sex: he to avoid paternity payments and protracted involvement; she to avoid the risk of having to raise a child without assistance from the father. But the relevant sexual instincts evolved long before the advent of efficient contraception and they will determine his need to 'score' just as much as her need for love and reassurance.

This variety-seeking urge in men, as against the reserve and selectivity displayed by women, is a sex difference that has been observed by anthropologists to be universal across human societies. Although cultures vary enormously with respect to overall levels of permissiveness versus puritanism, it is always the men who are more adventure-seeking in sex. Modern Western society probably comes as close as any to equalizing the opportunities for women to enjoy casual sex, and yet few seem disposed to capitalize on their freedom. Shere Hite (1976) found very few women among her highly liberated, educated, urban sample who gained any satisfaction from one-night stands, although many felt mystified and guilty because they thought they ought to. The vast majority still wanted sex with feeling, rather than sex for its own sake. For men, the reverse is frequently the case. Prostitutes report that a high proportion of their clientele are happily married men who are simply seeking a change from their staple sexual diet without risk of involvement. The proneness of men to sexual infidelity is sometimes alluded to by women with the expression 'Men are all the same'. That is not literally true, but they are on average different from women.

Homosexual men and women provide a very interesting test case of the desire for novelty in partners. These people are to some extent already outside social convention and therefore relatively free of sex-role prescriptions. They are also free of any concern about unwanted pregnancies. Surveys show that homosexual men tend to be highly promiscuous, with a turnover of partners that is between ten and one hundred times greater than that of their heterosexual peers, and they very seldom establish long-term relationships. Lesbian women, by contrast, tend to acquire long-term companions and enjoy physical sex mostly within the context of a loving relationship. It appears, then, that freed of the need to compromise with the proclivities of the opposite sex, the characteristic instincts of men and women cause their behaviour to diverge further in homosexual liaisons.

In laboratory animals, the male need for variety in partners is called the 'Coolidge Effect'. If a male rat is introduced to a female rat in a cage a remarkably high copulation rate will ensue initially. Then, progressively, the male will tire of the female, and even though there is no apparent change in her receptivity, he eventually reaches a point where he has little inclination to copulate. However, if at this point the original female is removed and a fresh one supplied, he is immediately restored to his former vigour. The same effect is seen even more strikingly in farm animals such as sheep and cattle. Rams and bulls have a definite aversion to repeating themselves with the same female, such that, for breeding purposes at least, it is unnecessary for the farmer to have more than one male stud to service all the female animals.

Symons (1979) points out that the term 'indiscriminate' has been applied most unfairly to male sexuality. Male animals do not choose their mates randomly; at the very least they identify and reject those that they have already had sex with. In the case of rams and bulls it is notoriously difficult to fool them that a female is unfamiliar. Attempts to disguise her by covering her face and body or masking her vaginal odours with other smells are usually unsuccessful. These male animals know where they have been and do not like going over the same ground again.

The origin of the term 'Coolidge Effect' is interesting to recount. The story goes that President and Mrs Coolidge were visiting a government farm one day, and shortly after arrival were taken off on separate tours. When Mrs Coolidge passed the chicken pens she paused to ask her guide how often the rooster could be expected to perform his duty each day. On being told that this would be dozens of times a day she was most impressed and said, 'Please tell that to the President.' When the President passed the chickens and was duly informed of the rooster's performance, according to the instructions of his wife, he inquired as to whether this was with the same hen each time. 'Oh no, Mr President, a different one each time' was the reply. The President nodded slowly and said, 'Tell that to Mrs Coolidge' (Bermant, 1976).

Although the Coolidge Effect is much diminished in intensity in the primates, and perhaps especially so in humans who have moral compunctions to deal with in addition, some substantial vestiges of it are nevertheless apparent. Before marriage it is usual for men to initiate intercourse at a fairly high frequency. After a few years of marriage, however, the husband's sexual appetite begins to wane and we quite often observe the emergence of what might be called the 'George and Mildred syndrome' with the now frustrated wife demanding more love-making than her 'tired' husband is able to supply. He, of course, is still perfectly capable of being aroused by his mistresses and other young girls, and if fortunate enough to secure an invitation to an orgy would have little difficulty in having sex with three or four different girls in the course of the evening's festivities. Many sex therapists (especially female ones) are inclined rather hastily to attribute the breakdown of sexual relations in a marriage to a prior breakdown in the personal aspects of the relationship. In fact, simple boredom with repetitive sex on the part of the husband may be an adequate explanation in a great many cases. In other words, the cessation of sexual activity in marriage can just as often be a cause of difficulties in the relationship as a consequence of them.

In this connection it is interesting to note that whereas involvement of a woman in extramarital sex is usually a signal that something is wrong with her marriage, there is no such association for men. In fact, there is some evidence that in men who have been married for a long time sexual experience outside the marriage contributes positively to marital happiness (Glass and Wright, 1977).

Intermale competition

The second thing to be derived from parental investment theory is that males will be driven by their polygamous desires to compete strongly for women. In many species vicious battles do occur between males for access to territory and females. Ideally, this should not be necessary, since there are an equal number of females and males and therefore enough to go around. The trouble is that the males of most species want more than their fair share of the females, or at least to impress and monopolize the most attractive of them (which usually means also the healthiest and best breeders among them). Presumably this is the primitive basis of the male interest in competitive sports like tennis and boxing, and perhaps even of the prevalence of wars and insurgency. Since those males with the most competitive instinct have in the past succeeded best in passing their genes to following generations, such traits have become highly developed in men.

The alternative solution is of course monogamy: the acceptance of the code according to which one male has exclusive access to one female for life, or at least for one breeding season. Some animals seem successfully to operate such a system, the advantage of competitiveness no doubt being offset by the enhanced chance that all the individuals will survive because peace prevails in the realm. But although there have been many attempts to encourage such a system in human society through institutions such as the Christian Church, they have at best been only partly successful.

Human males do seem to have a strong competitive urge built into their instinctual system. This is particularly dangerous to our species because our ingenuity has led to the development of weapons so lethal that they threaten our total extermination. In contrast, most other animals engage in power struggles which are used to establish a pecking order or dominance hierarchy without usually resulting in death to either of the combatants. It is this capacity and willingness to kill each other that has led man to recognize himself as probably the nastiest animal in creation. Actually though, we are no nastier in our nature than a great many other species; it is just that we have the power to wreak extraordinary destruction on our fellow man with no greater degree of malice.

So long as males compete with one another for access to as many females as possible, it is inevitable that the extent of their success will be uneven. Some males will gain control of several females, while others will necessarily miss out altogether. Studies of a great many species have confirmed this effect. Males show much greater variability in reproductive success than females, measured either by the number of partners copulated with or the number of offspring that result. While nearly all females manage to mate somehow or another, many male animals, especially those that are small and submissive, do not manage any heterosexual contact at all. The same principle applies to humans, and its relevance to the preponderance of male sex deviation has already been alluded to.

An important effect of this uneven success, also referred to in the previous chapter, is that males are the more modified sex. Darwin noted that males show a developmental spurt around the age of puberty, corresponding to an increase in testosterone production, which changes their physical and mental characteristics away from the childhood state far more radically than the changes that occur between girlhood and womanhood. This he believed was due to the fact that sexual selection operates more powerfully on males than on females. In birds, the process of male modification is carried to an extreme with the evolution of useless though spectacular decorations such as the peacock's tail. In mammals the effect is less striking because there is a greater tendency for sexually selected characteristics to be shared back with the other sex so as to maintain a relatively high degree of similarity between the two sexes of the same species. Nevertheless, the same principle applies, examples being the antlers of the stag and the mane of the lion. In humans it is also true that a grown man is more differentiated from children than are women, the most obvious physical adornment being the bearded chin.

The ladykillers

Evolutionary theory makes clear predictions about the type of male that is most likely to succeed in competition for females. We have already discussed the factors of sheer size and strength and of skill and ingenuity. But most disquieting from the point of view of predicting the future of the human species is the possibility that aggressiveness and psychopathy may go with breeding success in males.

One of the clinically established characteristics of psychopathy, which is almost exclusively a male condition, is an ability to impress and charm large numbers of women. Among the best-known psychopaths of recent times is Charles Manson, who managed to gather around him a 'family' of girls who lived with him in the Californian desert. So devoted to him were they that he was able to persuade some of them to perform the ghastly murder of the pregnant actress Sharon Tate on his behalf. Andreas Baader had such magnetic appeal to women that four of them, whom he had hardly met, helped him to escape from jail to start the German Red Army Faction, otherwise known as the Baader-Meinhof gang. Baader was a recognized psychopath well before his murderous nature acquired any political rationale or direction; in fact, the left-wing cause was largely contributed by Ulrike Meinhof. Before his imprisonment for arson, Baader had managed to move in upon a married couple and father a child with the wife, the husband apparently acquiescing in the project. This is reminiscent of Richard Wagner, another near-psychopath (however talented) who fathered children with Cosima Von Buhlow, without apparently losing the respect of her husband. In fact, Von Buhlow conducted the first night of *Tristan and Isolde* shortly after his wife had given birth to a child named Isolde, who was Wagner's daughter.

Brian Jones, the somewhat reckless and irresponsible guitarist of the Rolling Stones, left six illegitimate children, all by different women, before being found drowned in a swimming pool. Although one of the first pop musicians to wear makeup and wear outrageous androgenous clothing such as chiffon blouses and Ascot hats, 'he carried such an aura of street-guerrilla aggressiveness that no one would dare suggest to his face that he looked less than totally masculine' (T. Sanchez in *Playboy*, August 1979). Charles Sobhraj, a Franco-Vietnamese man who preyed on idealistic travellers on the Eastern hippy-trail, and probably killed about twelve of them, was also described as charismatic in personality and highly attractive to women.

Whatever the extent of their callousness and brutality, there is something in the primitive hyper-masculinity of these men which seems to have a magnetic appeal to *some* women. It is also seen in the willingness of nurses and female prison visitors to fall in love with and marry impossible psychopaths and convicted multiple murderers. The genetic implications of this phenomenon are quite frightening. If psychopaths are exceptionally successful with women, then the genes for psychopathy will gain ground relative to those of milder men. It seems they already have to some extent, for cold-bloodedness, aggressiveness and criminality are clearly more characteristic of men than women. Perhaps it is only the execution, castration, and incarceration of extreme psychopaths throughout history, combined with the tendency of many to eliminate themselves in the course of their desperate adventure-seeking, that has prevented men from becoming progressively more dangerous over recent centuries.

Competition for mates or mating?

Although it is certain that inter-male competition for reproductive advantage is a strong contributor to the double standard, this advantage can be achieved in different ways. If we compare the sexual behaviour of two of our closest relatives, the chimpanzee and the gorilla, we find that their sexual habits and living arrangements are very different one from another, and it is interesting to consider which one of them provides the better model for man.

Chimpanzees live in loose communities with roughly equal numbers of males and females in the troop, and with the males continually vying for female favour. Gorillas, on the other hand, approximate more to a harem system. A dominant male tolerates only one or two young males, who are usually his sons, and who will take over the leadership when he dies; the other young males leave the group and wander alone until they are able to find their own mates. These different living styles are believed to be connected with the extent of the size difference between the male and female of the species. Male chimpanzees are only slightly bigger than the females, but the male gorilla is about twice as big as the female and his size and strength appear to be instrumental in establishing ownership of a harem.

> Some adult male gorillas have a large number of females in their group, while others wander alone and so have none. Unless these loners acquire females from neighbouring groups, and manage to keep their females with them, they will have no chance of producing offspring. Not surprisingly, therefore, the males compete quite openly for females. They usually do this by chest beating, ground thumping and foliage slapping, and no-one gets hurt. Sometimes, however, fierce fights do break out; the aftermath of these can be blood bespattered, flattened vegetation over many square yards, with the antagonists bearing wounds for days, sometimes weeks afterwards. . . . When fights between males are so severe a large body size is up to a point an advantage; the winner is likely to be the larger animal . . . (Harcourt and Stewart, 1977).

What is to stop the chimpanzee males from fighting over females and so gaining in size relative to them? The dominant males frequently try to prevent the others from mating but they are much more gentlemanly about it than the gorilla. One possible reason is that by fighting with another male the chimpanzee runs the risk that a third male might sneak in and score with the female while the combatants are occupied. After all, there are plenty of males around in the chimpanzee troop who are prepared to take their opportunities as they find them. The gorilla does not have this problem since he seldom has to contend with more than one intruder at a time.

Another factor that probably reduces the extent of hostility among chimpanzee males is that the rivals are usually related to each other (and thus sharing some of their genes anyway). Chimpanzees are fairly unusual among primates in that the males do not leave the group in which they were born in order to find unrelated females with whom to breed; rather they remain in the community amongst their relatives. It is the adolescent female chimpanzee who counters the danger of inbreeding by leaving the home troop to join a neighbouring one. The advantage of this system, it seems, is that the males are less inclined to injure each other in jousting for female favour. Gorilla males who fall into conflict are usually unrelated and therefore have no such inhibitions.

Certain other physical and social differences between chimps and gorillas also make sense within this context. The female chimp in oestrus has a much more conspicuous pink backside than the female gorilla, whose sexual invitation is considerably more subtle. Harcourt and Stewart suppose that this is because the female chimp finds it profitable to signal her readiness over some distance, so as to attract a selection of males from which to choose. The female gorilla is, of course, already committed to a particular male with whom she lives intimately, so it would be unnecessary (not to say indiscreet) to advertise her charms more widely. Similarly, the chimp's oestrus lasts about ten days, giving several boys a chance, whereas the female gorilla is on heat only for one or two days, which is no doubt enough considering her husband has several wives to service as well.

Then there are differences in the vigour of the male courtship display. The male chimp competes for the favour of the oestrus female by flamboyantly erecting his hair and his penis and energetically waving the branch of a tree. Only if this ritual is impressively performed will the female show the sexual willingness upon which he is dependent. Once he has his females in the group, the male gorilla has no need to pursue such formalities in order to gain their cooperation. The female is virtually ignored until copulation takes place, and she has very little choice when it does.

Chimps do not always need to perform their courtship displays. Sometimes they are able to persuade a female to go off with them alone on a kind of honeymoon, and quite a high proportion of conceptions occur in this way. At other times, a kind of 'gang bang' takes place, with several males gathering around one female and mounting her in rapid succession. In the latter situation it pays to be highly virile, and it is perhaps for this reason that the male chimpanzee under pressure from others can mount, thrust, ejaculate and dismount all within the space of about seven seconds. Gorilla copulations are much more leisurely, lasting usually for a couple of minutes and sometimes up to a quarter of an hour.

The relative size of the male equipment is also understandable in light of the above. Although the gorilla is very much bigger than the chimpanzee overall, his testicles are only one-sixth the size. Since he does not share his wives with other males he has no need to compete with them as regards the quantity and concentration of sperm provided in one ejaculation. By contrast, when male chimpanzees engage in group sex with the same female, those that contribute the most powerful load of sperm will be at an advantage in that they are more likely to be responsible for any fertilization that takes place.

The interesting thing about this comparison of chimpanzees and gorillas is that, although closely related to each other, these two types of ape have arrived at very different systems of courtship, sexual behaviour and social living. In both cases, the males engage in more reproductive competition than do the females, but the manner of their competition is quite distinctive. Also, it is clear that the physical characteristics of both males and females are logically integrated with the sexual habits and life styles of the respective species.

Which of these two apes comes closest to the human pattern in terms of bodily characteristics and mating patterns? The most objective comparison is probably the male/female size discrepancy, and with respect to this criterion we fall between the other two apes, but closer to chimpanzees than gorillas. Men are clearly bigger than women but the size difference is not staggering (about fifteen per cent of body weight) and there is a great deal of overlap. Likewise the testicles of a man, relative to his overall body weight, are also intermediate in size between the chimpanzee and the gorilla, although the man's penis is larger than either. On these bases, we might expect our sexual and marriage habits to fall somewhere between the gorilla and the chimpanzee but perhaps closer to the latter.

In fact, human mating behaviour does seem to involve aspects of both chimpanzee and gorilla behaviour. Certainly, both forms have been observed in various times, places and cultural groups. Men do fight over women to some extent; duelling is a well-recognized historical practice, and dance-hall fights are still quite common. Considerable injury may be inflicted, though it is difficult to tell to what extent this is due to our cleverness in developing weapons far more lethal than our own fists and teeth. Perhaps on balance we are more gorilla-like as regards the potential for aggression, and consistent with this is the fact that male rivals are seldom blood relatives. The other respect in which we behave like gorillas is that we do establish relatively undisputed ownership over women (through the institution of marriage) that is not related to a hierarchy position, and the result is that we no longer need to impress them with politeness or physical prowess before taking sexual gratification. Many societies that frown on rape do not concede that a husband can be guilty of raping his own wife.

It is difficult to make appropriate comparisons as regards the visual provocativeness of women because we cannot see a lady ape through the eyes of her suitor. However, it is consistent that their physical invitations usually become much less explicit after marriage, either because they 'let themselves go' or because they are deliberately protecting their husband's privilege by contriving to be unattractive to other men.

In other respects we behave more like chimpanzees. Bachelor men engage in elaborate courtship rituals like dancing, buying flowers and dinner, demonstrating their skill on the sports field, and driving impressive cars. Married men do the same, although seldom for the benefit of their wives. Bachelor girls will signal their attributes and availability quite unambiguously and in many directions at once with the aid of cosmetics and striking clothes designed to set off their natural biological signals. Oestrus in the human female occupies an even greater proportion of the menstrual cycle than it does with chimpanzees; in fact, a woman is receptive to a varying extent at all points in her cycle. Sex between one woman and several men (either with or without her consent) is by no means unknown. With respect to genital size men are more in the class of chimpanzees than gorillas, though their staying power in intercourse is usually more like the latter. It seems, then, that the sexual and social behaviour of man falls somewhere between that of the chimpanzee and the gorilla, or at least is so variable that it straddles the range displayed by our two Simian cousins.

Rapists of the jungle

The behaviour of chimpanzees and gorillas does not exhaust the possible analogies between man and other primates. Orangutans are different again. They move around as solitary individuals a great deal of the time, but when a male encounters a female, sex is quite likely to occur. Sometimes, like the chimpanzee, the female appears to engage voluntarily in a temporary consortship with a particular male. But at other times she is taken quite forcefully in a manner that looks to us like rape. When first assaulted the female orangutan will struggle and emit distress signals, but once she is pinned down convincingly and thrusting has begun she appears to submit reasonably amicably to her fate (Nadler, 1977).

Interestingly, the idea that many women who initially protest will, if properly subdued and stimulated, eventually get around to liking what they get, corresponds to a very popular male fantasy. It is also the theme of a great many pornographic movie sequences. Women are also reputed to have fantasies concerning rape. Some people claim this as evidence for a buried desire to be raped, but in most cases such fantasies probably function as a way of preparing to cope emotionally with the possibility of a genuinely dreaded event. Of course, even if it were true that some women who are raped end up by deriving some form of sexual satisfaction from it, this does not make it morally defensible within a human context. We would still regard the woman's freedom of choice as sacrosanct at the moment she is first propositioned. The truth is that most women want to be 'swept off their feet', figuratively if not literally, by a strong, dominant man, but they also want the time, the place and the particular male all to be right before this happens.

Keeping the harem happy

Another primate group that provides an interesting comparison with human behaviour is that of the baboon. According to Dunbar (1978), the gelada baboon that inhabits the remote northern plateau of Ethiopia moves about in large herds which may be analysed into a number of harems. Each harem is made up of a single adult male and around three to five females and their offspring, although their size varies considerably. The smallest unit observed was one male and one female, while the largest contained twenty-eight individuals, including a dozen adult females.

Apparently, the main determinant of the number of females a male can maintain in his harem is his ability to pay them all sufficient attention. This he does by grooming them periodically. If the male does not groom his wives with sufficient regularity there is a good chance that they will desert him and join the harem of a challenging male. The challenge usually takes the form of an attack from a younger male who has not yet acquired a harem. Even if the attacker is at an advantage in terms of fighting strength and skill and is able to win the fight, the harem may still choose to remain loyal to the older male if he has maintained a sufficiently good social communication with them. But if so much as one of his wives chooses to desert in favour of the usurper, there will probably be a mass exodus in that direction because all the females in the harem will have developed strong social bonding by grooming each other when their husband is otherwise engaged. The dilemma of the male, then, is whether he can risk adding another female to his harem and still groom all of them sufficiently to keep them happy. If he fails, and the harem deserts in favour of a younger, less committed male, he will probably never manage to rebuild a harem.

There is a definite parallel here with certain human societies in which a man is culturally permitted to take on extra wives so long as he can provide for them. In our society, the man who expects to maintain a mistress as well as a wife, or several girlfriends simultaneously, will find he has to devote a great deal of time to each of them or they will shift their allegiance to another man. Like Falstaff and the Merry Wives of Windsor, the man who overplays his hand and tries to score with too many women simultaneously is likely to end up all alone and in the drink.

Competition after copulation

Not all of the competition between males for siring offspring with the females occurs before the act of intercourse. In fact, there is an even greater variety of competitive tricks in the animal world that are played out after insemination has taken place. Many insects employ the technique of plugging the female's genital tract by a coagulation of secretions after their sperm has been deposited. This has traditionally been supposed to prevent the sperm from leaking out again afterwards, but recently it has been realized that it also serves the purpose of preventing other males from adding their sperm to the pool. Other insects use themselves as a genital plug, staying *in situ* and riding tandem for several hours after depositing their sperm and thus preventing any displacement of it by other males. The poor male of one species of fly has to suffer being eaten alive by his wife, but his genitals remain in place to act as a mating plug, so that his mark is almost certain to be left in the offspring. Mating plugs also occur in a number of mammals including hedgehogs, bats and rats. Chastity belts and infibulation (stitching the entrance to the vagina) would seem to be human equivalents, though they require equipment that is not supplied in our biology.

Male mice have the rather heartless capacity to produce an odor that will cause a pregnant female to abort and thus become available for reinsemination, while langurs and lions routinely kill all the infants in a troop that they have conquered by driving off the resident males; they then promptly inseminate the females again. It would be wrong to describe this activity as inhuman since it has been practised by primitive peoples such as the Amazonian

Indians as well as by the more 'civilized' races when they become 'carried away' in wartime.

The males in some species, including the chimpanzees discussed above, are discourteous enough to attempt a takeover of the female during the act of copulation itself, sometimes making concerted group attacks on the pre-occupied male. This too is not unknown in human society, as when gangs of juvenile rapists attack couples making love in parked cars at night, and indulge in a 'gang bang'. The result is that couples throughout the animal world often seek privacy for their sexual encounters. Many apes and monkeys, like humans, disappear for secret honeymoons away from the rest of the social group to engage in their most intense periods of sexual intimacy. After returning from honeymoon, many men like to restrict their wife's contact with other men in order to avoid the dread fate of cuckoldry, and so they attempt to persuade them with a variable degree of success that their proper place is in the home.

Female coyness

Complementary to the male tendency to be lustful and promiscuous is the female pattern of coyness and reserve in matters of sex and courtship. It is characteristic of the vast majority of animal species that during the mating season the male is readily aroused by the sight of the female but is required to go through a specific ritual, sometimes quite elaborate, before being allowed to proceed to copulation. Similarly, in human societies female caution and modesty appear to be fairly universal. It has often puzzled anthropologists by appearing in cultures which adopt a very permissive attitude towards sex and in which the adults accept sex play in both male and female children as perfectly natural and healthy.

In Western society, the pattern of female reserve looks

set to survive the influence of the current feminist movement with consummate ease. An illustrative study is that of Peplau, Rubin and Hill (1977) whose subjects were 231 dating couples in the area of Boston, Massachusetts. In the majority of cases these couples were already sleeping together on a more or less regular basis, but it was fairly clear that it had been the woman who had exerted control over when intercourse should first take place. Various characteristics of the female partner were found to predict the timing of first intercourse quite strongly, while the characteristics of the man were largely irrelevant. No doubt his interest in the project could be taken for granted.

The girl's religion, for example, was a strong predictor; 27 per cent of the Catholic women had resisted premarital intercourse altogether, compared with 16 per cent of Jewish women and only 2 per cent of Protestants. Religious affiliation in the man had no such effect. The girl's previous experience was also an important factor. If she was already sexually experienced, intercourse took place in an average of two months regardless of whether or not the man was experienced. If, however, the woman was a virgin at the beginning of the relationship, the experience of the man did assume some importance. If the man was experienced she would take an average of six months to get around to having intercourse; if he was inexperienced it would take twice that long. Presumably this difference reflects the libido and seductive skill of the man.

If the couples were not yet having intercourse, this was nearly always because the girl did not yet feel ready. Very few couples gave reluctance on the part of the man as a reason. It was also interesting to note that if the girl had not yet 'given herself' sexually she tended to hold more power within the relationship in terms of deciding the couple's activities. This would suggest that she had also exercised a positive decision to the effect that intercourse would not be on the agenda. The female resistance to sex cannot then be interpreted as a lack of assertiveness, as some feminists have suggested. Consciousness raising and assertiveness training of women will not turn them into derepressed men, it will only make them say 'no' more decisively when they feel reluctant to engage in sex.

There was a time when virginity was held in such esteem that a girl was afraid her boyfriend would lose all respect for her if she allowed him to 'go all the way'. Since he would be bound to boast of his conquest to other males this would probably diminish her marriage prospects all round. Today, premarital sex has become so commonplace that a girl often fears she will lose her boyfriend if she does not permit intercourse within a reasonable period of time. Christensen and Gregg (1970) found that nearly a quarter of their sample of college women had 'given in' to their first sexual experience without feeling desire on their own part. Rather, they had yielded either to force or from some sense of obligation to their boyfriend. Less than three per cent of men questioned gave answers that could be classified the same way. Likewise, when Bardwick (1971) asked college women why they had first engaged in intercourse, typical answers were along the following lines:

> 'Well a great strain not to. Fairly reluctant for a while, but then I realized it had become a great big thing in the relationship and it would disintegrate the relationship . . . I wanted to also.'
> 'He'd leave me if I didn't sleep with him.'
> 'Mostly to see my boyfriend's enjoyment.'
> 'I gave in to Sidney because I was so lonely.'

Very few of the girls that Bardwick interviewed said they had started having sex because they wanted to. Usually it was regarded as a price that had to be paid to keep a relationship going, or it was an attempt to prove to their boyfriend that they loved him. Despite the great increase in permissivness in our society over the last few decades, the motives and concerns of the two sexes remain quite different.

The characteristically female trait of coyness makes its appearance very early in life. At around the age of one or two, little girls are often observed to hide their eyes from strangers in a manner that seems to express something halfway between embarrassment and flirtation. This gesture is more typical when the stranger is an adult man other than the father, and it is much less common in little boys. That it is an instinctual behaviour pattern rather than one that is learned by imitation is suggested by the fact that it also occurs in girls who are born blind and would therefore have no chance to learn it off other girls or adult women. The seductive element in this gesture has led some ethologists to interpret it as a ritual invitation to chase (Eibl-Eibesfeldt, 1971). They note that a very similar sequence of making eye-contact with a man and then modestly averting the gaze downwards and away is displayed by adult women at cocktail parties and other sexually charged situations. This sequence is perceived by men as appropriate and appealing, whereas brazen staring and overly explicit invitations are seen as unfeminine and are apt to be counterproductive. In Gilbertian terms, when a lady is propositioned, even by a man she definitely desires, 'a certain show of reluctance will not be misplaced'. 'Playing hard to get' is not just a useful reproductive strategy for woman, it also appears to be a turn-on to men.

Caution in courtship may serve a variety of biological functions. One is species identification. Male fireflies, the great 'flashers' of the insect world, emit a special sequence of flashes which is read by the female like morse code. She will then respond with her own flashes only to a male of her own species, ignoring all other flash patterns. The courtship ritual may also allow the female to recognize an amorous approach from a male so that she can distinguish it from a more typical aggressive approach or from her own prey. When a male trapdoor spider performs his courtship dance outside the lair of a female it is vital that he get it right, for if he does not she is likely to leap out and devour him.

In higher mammals, the reserve shown by females to advances from unfamiliar males is best understood within the terms of parental investment theory as a screening process by which only certain males are selected to have sexual favours conferred upon them. The two classes of males that are likely to be most acceptable are those that comprise the 'fittest' breeding material, and those who are most available and willing to assist with future infant care. Since males are inclined to be deceptive and misrepresent themselves on both scores, it is in the interest of the female to put them through a protracted courtship procedure in order that their credentials may be thoroughly scrutinized before sexual privileges are granted. This is her best protection against having her precious eggs fertilized by weeds in padded football jerseys, mental retardates who have learned a few lines of Shakespeare, and bigamists with several other families to support. It will also give her an opportunity to see if some of the pair-bonding effects, whether they be described as imprinting or falling in love, will take place between her and the male in question, for if they do there is an increased likelihood of receiving continued interest and support from him.

Of course, a female who is fairly unattractive relative to her peers, and therefore not herself in great demand, may not be in a position to satisfy all these criteria simultaneously. She may, in fact, have to make a choice between responding to the promiscuous overtures of prime quality males who are philandering widely and perhaps are committed elsewhere, or settling for a dull but secure attachment to a less impressive breeder who offers reliable support. Many women are faced with the dilemma between enjoying casual sex with attractive playboys and high status males on the one hand, and permanent involvement with a boring though steady individual who promises to be a good provider on the other. Many women will do both – in that order. Only a few will opt for the other alternative, which is to marry for emotional and economic security while at the same time indulging in sexual thrills (and perhaps also impregnation) outside of the marriage.

Biological imperatives and human ethics

It has been the intention of this chapter to show that much of the behaviour that goes to make up the eternal 'sex war' is due to the fact that males and females have evolved partly discrepant sex instincts. These differences are much the same for animals and for humans in different cultures, and they can be seen indirectly to serve the reproductive success of males and females. It seems too much of a coincidence that 'society' should arbitrarily have arrived at these 'stereotypes' for male and female behaviour. More likely, such gender-linked behaviour patterns have evolved because successful mating strategies have been enshrined in the genes and are manifested as different instincts and emotions, mediated through the brain and primed by the endocrine system. What is socially learned, then, is not a randomly chosen set of (male-oriented) prescriptions for sex-role behaviour, but realistic expectations based on accurate observation of the characteristic inclinations of men and women that are part of their fundamental nature. Cultural pressures may reinforce or counteract these instincts to a greater or lesser degree but are nnever alone responsible for determining sex-role behaviour.

It may be necessary to point out to the reader who is not used to thinking in sociobiological terms that the instincts which serve gene survival operate unconsciously to a large degree. While the individual's behaviour is accompanied by strong feelings and needs, he or she will not normally be aware of the genetic advantages of the chosen strategy. People are in fact given to ingenious moral and intellectual rationalizations for their behaviour that are socially learned and superficially plausible. 'I don't believe in having sex before marriage;' 'I'm saving myself for the one I love;' 'My wife doesn't understand me;' 'You only live once, so let's make the most of it.' It should not be difficult to identify the gender of these deeply philosophical comments, and it is clear that they are useful not just for explaining our inclinations to ourselves but also for persuading persons of the opposite sex to cooperate in accommodating them.

Evolutionary theories of the kind espoused here are totally amoral. They are not concerned with the question of whether certain human tendencies are right or wrong; they are merely concerned with describing them and explaining their dynamic origins. We shall discuss the extent to which it is possible to override human nature by forces such as 'civilization', politics and religion in the final chapter. In the meantime it may be provocative to suggest that, as it applies to sex and marriage, the feminist philosophy, like some aspects of Christianity, could be construed as a moral system intended to promote female interests such a monogamy and paternal assistance with child-rearing. As such, it would have no greater claim to being a revelation of 'natural justice' than the

'Playboy philosophy', which is probably an equivalent statement of the male position. If justice has any meaning at all in this context it would be concerned with equalizing the extent to which members of each gender are required to bend their natural inclination in order to arrive at a peaceful compromise for social living. Exactly where this balance should be struck is at the heart of the current debate between chauvinists and feminists.

4
Attraction and Arousal

The evolutionary approach to the understanding of human sexuality provides us with a powerful framework within which to examine sexual preferences and also the differences between men and women in their mechanisms of arousal. It has often been asserted that 'One man's meat is another man's poison,' and that 'Beauty is in the eye of the beholder.' Such aphorisms imply that partner choice is a somewhat random and unfathomable process. Other commentators have pointed out major cultural differences in what is perceived as attractive and noted that in our own society standards of make-up, dress, and even body shape may change over time according to prevailing fashion. Therefore, they conclude that social forces are in some way responsible for defining what turns us on in a potential partner.

This chapter attempts to show that despite this bewildering variability in the criteria of beauty, there are powerful underlying rules which direct our search for sexual partners and which control the timing and extent of our arousal to them. Once again, these rules of behaviour may be traced to biological instincts that in the final analysis are related to reproductive fitness.

Criteria of beauty

Desmond Morris (1977) has described some of the amazing differences between cultures as regards what is thought to constitute physical beauty.

> Beautiful girls persist in changing shape as epoch succeeds epoch, or as the girl-watcher travels from society to society. In every instance there are fixed ideals which are hotly defended. To one culture it is vitally important that a girl should be extremely plump; to another it is essential that she should be slender and willowy; to yet another she must have an hour-glass shape with a tiny waist. As for the face, there is a whole variety of preferred proportions with almost every feature subject to different 'beauty rules' in different regions and phases of history. Straight, pointed noses and small snub noses; blue eyes or dark eyes; fleshy lips or petite lips; each has its followers.

Morris goes on to note that although the Miss Universe contest pretends to be cross-cultural, judgements are in fact made largely in accord with Western ideals of beauty. If the competition is won by a black or oriental girl it is because her figure and face are unusually Western looking, and she would probably not be judged a beauty by her own people. 'Girls coming from cultures where protruding buttocks, an elongated clitoris, or unusually large labia are the most prized features of local beauty need not apply. They would never reach the semi-finals.'

This cultural variability in what is seen as attractive is often taken by anthropologists to mean that there are no biologically determined standards of attractiveness – that beauty is entirely culturally determined. Such a view is mistaken. The differences that do exist from one culture to another have been emphasized by researchers and writers because they are of greater interest than the similarities. Anthropologists have gone into other cultures with the prime intention of documenting ways in which they differ from our own. The result is that they often pass over or fail to notice the universals of human behaviour.

Such cultural variability that does occur can be explained in two ways. One is in terms of associated characteristics. In some cultures fatness is a prized characteristic, and this may be because only the rich and people of high status can afford enough food and drink to become fat. In our culture, where obesity is more of a problem than starvation, thinness is more highly valued. With us, a suntan is often thought to be attractive. This is partly because it looks healthier (people who are ill frequently lose colour in the face), but it may also be because of the prestige associated with tropical holidays. People with the money and freedom to trip off to the Mediterranean or Caribbean inspire jealousy in others and the suntan thus acquires conditioned value. Whereas we in the West admire the firm breasts of young women, there are some African tribes which apparently place a high value on drooping 'pendulous' breasts. Closer analysis reveals that these tribes also place a high value on age, so the drooping breasts are probably admired because of their connection with maturity in women. Other things being equal, the biological tendency is to be attracted to younger breasts since they are in prime condition to serve a reproductive (suckling) function.

In evaluating the sources of cultural variability in attractiveness it is important to note the distinction between sexual attraction and aesthetic beauty. Morris argues that the former is largely biologically determined according to the principles of reproductive significance, while the latter is a culturally variable spin-off of the very human need for, and enjoyment of, classification. Art, according to Morris, is primarily a product of a 'taxophilic urge' which causes us to compare and grade stimuli in all areas of interest. No doubt we are genetically primed to enjoy flowers and trees in general, but we humans have an additional tendency to assess the variations among them according to certain criteria, so that certain flowers and certain trees are seen as the most beautiful exemplars of their class.

Standards of aesthetic beauty vary from time to time and place to place for a number of strange reasons. Fashions in beauty may change like fashions in clothing because of economic conditions. In the case of women's clothing it appears that sexually provocative presentations are more common in times of economic security. Thus, over the last eighty years skirt lengths have been a fairly reliable economic barometer. Although it is not clear why this should be so, it is likely that people have the time and inclination to play when things are going well – and this includes sexual play. When times are tough, and in poor societies, an austere puritanism is more likely to prevail. No doubt the same rule applies to fashion in faces and bodies. When there is a great deal of poverty and privation, practical virtues such as strength and plumpness are likely to be most coveted and social graces such as courage and altruism will be emphasized. In conditions of plenty, the more luxurious and 'effete' attributes such as fine features, slimness and playfulness will come into vogue.

This is just one example of a transient cultural factor that might influence local and contemporary standards of beauty. Another is the simple need for novelty that seems to determine the restless nature of teenage pop cultures as well as clothing fashions. In this context models are of paramount importance. Particular individuals (usually singers, film stars or sportsmen) have an important impact not only on hair styles and clothing fashions but also on standards of facial and physical beauty. Jane

Russell helped create an interest in large breasts besides reflecting a cultural trend, and the combined talents of Mark Spitz and Robert Redford in the mid-seventies led to a remarkable increase in the number of young men sporting moustaches.

The thing to note about all these variations is that they are relatively superficial. There are, underlying them, some very important general rules of sexual attraction and arousal.

Gender differences as the basis of sexual attraction

Although it may seem too obvious to be worthy of mention, sexual attraction is based primarily on the existence of physical differences between men and women. Many of these physical differences have come about because of the specialized roles of men and women. Because men have evolved as hunters they are taller, more muscular, have proportionately larger feet (for fast running), larger hands and forearms (for weapon wielding), bigger chests (for stamina) and stronger skulls and jaws (for protection). Because women are specialized for child-rearing they have a broader pelvis, larger breasts, a greater gap between the upper thighs, prominent buttocks, a rotating gait, and a higher proportion of body fat. All of these gender identifying features are likely to be seen as 'sexy' by persons of the opposite sex, and within limits their exaggeration will be perceived as even sexier. Thus tall men are especially prized and so are large-breasted women.

There are also certain odour differences between men and women which may still play a role in sexual attractiveness, albeit a somewhat diminished one compared with other mammalian species. The odours are produced by special scent glands located particularly in the hairy pubic and underarm areas. Although it is sometimes said that the hair in these areas had been retained by the human ape for lubrication, their more important function is probably to trap these sexually attractive scents. Unfortunately, by wearing clothes in addition we often trap the odours for too long and they become stale and offensive; thus politeness requires us to wash frequently. Many people go further and use deodorants or shave their underarm area.

If smells do function as sexual attractants in humans they probably have more effect on women than men – at least in the absence of commercial perfumes. Men naturally produce stronger underarm smells and women are more sensitive to them. However, the visual mode is more important in the early stages of sexual attraction and, as will be discussed later, men are more aroused by the sight of a woman than vice versa.

The evolutionary development of this male voyeurism is interesting. In many other primates, the main stimulus to sexual activity is the presentation by the female of a reddened, swollen bottom to the chosen male, who can usually be counted upon to do his duty. But since the higher primates, especially homo sapiens, spend much of their time in face-to-face communication, it became advantageous for the female to mimic her rear-side invitation signals on the front side. This is seen most obviously in the female gelada baboon who has an unmistakable copy of her rump-display on her breast. The nipples have come to lie close together so as to resemble the genital labia and are surrounded by pink, naked skin similar in colour pattern to the backside. Even more impressive is the fact that the breast display varies in colour intensity in synchrony with the rump display. In this way, the female gelada can attract male interest at appropriate times from the front while sitting down.

Morris (1971) argues that the human female has evolved

similar genital echoes. Human breasts are large, round and pink and often have a mysterious cleavage between them which is of special erotic significance. The lips also echo the instinctual genital invitation to some extent. The word 'labia' actually means lips. Female lips are more everted (i.e. larger) than those of men and they have often been reddened and moistened in order to enhance their erotic appeal by emphasizing the genital association. Other areas of the body which are also smooth, pink, rounded, and arranged in pairs may also have sexual connotation. Photographers exploit this when they have their model raise a shoulder against one cheek or bring their knees up to their chin.

It is difficult to tell at what point the biological priming for response to the genital display gives way to learned associations. 'Echoes' of the female rump such as shoulders, knees and lips may come about as a result of simple generalization, or they may be conditioned associations (a rather automatic linking of the two stimuli because they occur frequently in quick succession). Again, they may occur because of more complex symbolic associations. Although the latter level has been given emphasis by Freud and the psychoanalysts, it may be entirely unnecessary. The various female body parts are probably reminiscent enough for direct stimulus generalization to operate. However, it is very difficult in a context such as this to distinguish learning processes from developments which are genetically almost inevitable. The old dichotomy between instinct and learning will probably turn out to be as fruitless as that of free will versus determinism. Most of our behaviour results from learning that occurs within the confines of very powerful genetic priming.

Age and attractiveness

A second major principle of attractiveness and partner

choice concerns the importance of youthfulness in the female. In the animal world it is quite usual for the males to copulate with females who are younger than themselves; age is not an important factor with the male so long as he can defend his territory and female consorts against younger usurpers. In human society too, it is quite clear than men prefer women younger than themselves. They marry women who are on average a few years younger, while their mistresses and girl-friends are usually younger still. If asked to judge sexual attractiveness in the abstract, men of all ages choose girls around the age of sixteen to twenty-five their ideal. No such prejudice is shown by women.

One rather superficial explanation of this pattern is that girls mature earlier than boys. While it may be true that puberty is reached a year or two earlier in girls, this does not help to explain why the male preference for younger partners carries on throughout life. The real reason is that the reproductive fitness of women declines rapidly with age. Females produce their best babies when young – in humans between the ages of about eighteen and thirty-two. Since, as we noticed earlier, the female's eggs are present in her body throughout life they are susceptible to deterioration as a result of chemical interference or radiation. The result is a very much increased risk of genetic anomalies, the best-known of which is Down's Syndrome (Mongolism). Male sperm is generated freshly throughout the lifetime, and that produced in old age is just as good as that produced when young, even though it may not be emitted with such frequency and concentration. Thus it is that males of all ages are competing for prime-time females, and frequently the old ones are better organised for this competition, having accumulated territory, social position or just plain experience.

From the female point of view, choosing older males gives them a better chance to assess breeding quality since the various men can be judged by their achievements over the course of a longer lifetime. Distinguished men like Charles Chaplin and Aristotle Onassis have little difficulty in attracting wives. Although their physical looks may diminish to some extent, their achievements in life remain as testimony to their desirability as breeding material. Some older women like Elizabeth Taylor and Jacqueline Onassis can retain their fascination because of exceptional personal qualities (and wealth?) but it is much more of an uphill battle for a woman to maintain desirability in the face of fading physical attractiveness than it is for a man. Confirmation of this fact comes from research which shows that women who had been beautiful when young tend to be less happy with their lot in middle age than women who had always been plain; for men there is no relationship of any kind between physical attractiveness and happiness over the course of a lifetime. Presumably, the women who started life as beautiful and desirable suffer more of a sense of loss as they get older than those whose happiness was never built upon such a foundation (Wilson and Nias, 1976). One wonders to what extent this problem is basic to the psychiatric difficulties faced by many beautiful actresses as they begin to pass their prime. Jean Harlow, Judy Garland, Marilyn Monroe and Rita Hayworth are among the Hollywood stars that come to mind in this connection.

In the Blake Edwards film *10*, one of Dudley Moore's peer-age female admirers is depressed by having observed his interest in the youthful Bo Derek and asks the barman what is fair about the fact that as a woman gets older she becomes less desirable while the man who begins to go grey becomes progressively distinguished. 'Lady,' replies the barman, 'there is nothing fair about that.' The problem that most egalitarian movements have to face at some time or another (in this case certain branches of feminism) is the fact that nature has no sense of justice; it is concerned only with survival. The preference of men for young girls is no superficial response to advertising and media propaganda, it is a result of genetic programming that will not be eliminated by any amount of preaching and idealism. At best it can be recognized for what it is and brought under moral control.

Because young women are at such a premium and the young men have to compete not just with each other but with older men as well, there arises a problem of loneliness and frustration in older women and in younger men. Computer dating companies and marriage bureaux inevitably find themselves with a surplus of these two groups on their books and have great difficulty in suiting them. Some find it necessary to refuse membership to people in these categories or deter them with higher charges. A solution that has been proposed by some recent writers (for example Sandy Fawkes in her book *In Praise of Younger Men*) is for these two groups to get together. The suggestion is that older women should use their newly acquired liberty to take the initiative in seducing younger men. This should not prove too difficult a task because younger men usually have an excess of libido that is all too often wasted on masturbation. Therefore, getting together with younger men may be a reasonable adaptation for some older women. It may also be a lot of fun for the young men concerned. It is unlikely to be a completely satisfactory solution, however, because the older women are probably more in need of stable companionship while the young men are primarily seeking sexual thrills. While these women may be able to hold the interest of their young man with their maturity and experience for a while, they will risk being hurt when he moves on to a younger woman, as he is very likely to do sooner or later. Provided the women are braced for this almost inevitable contingency both parties may find it a rewarding experience.

The face and eyes

Despite the popular assertion that 'You don't look at the mantelpiece . . . ' the face is of special importance to human attractiveness. This is partly because it is the most expressive part of the body and therefore a betrayer of personality and temperament – for better or worse. The skin tone is also a key indicator of age and health. As people get older their facial skin begins to wrinkle and sag, this decline being more merciless on women, who have the more delicate skin surface to begin with. Since in some parts of the world, particularly the United States, many people undergo cosmetic surgery to remove wrinkles from the face, the skin on the neck or the hands is actually a more reliable indicator of age. When people are in good health their skin is fairly warm in colour and clear of blemishes, though of course the variation is more obvious in the white races.

While it is difficult to make rules about what constitutes a beautiful face, symmetry and the absence of abnormal characteristics are probably key features. A clue to this is the finding of Sir Francis Galton, back in Victorian days, that if the photographs of a motley collection of people (all of the same sex) are superimposed one on top of another so as to form a composite, the final portrait is usually judged as beautiful. Presumably this is because the peculiar characteristics of each individual are cancelled by the averaging process so as to leave a fairly 'ideal' face devoid of any striking abnormalities and blemishes. However, this is not to say that any special identifying characteristics we might have are necessarily a liability.

Many early analysts of human attachment behaviour, such as Binet and Krafft-Ebing, have pointed out that when somebody falls in love with us it is usually our peculiarities that they imprint upon – the things that make us unique as individuals. This observation, if it is true, should make people think twice before submitting their face to the plastic surgeon's knife for remodelling. People who acquire new faces in this way may traumatize their family and loved ones as well as themselves because they interfere with the fetishistic elements of human bonding.

Another key indicator of physical health is whiteness of eyes. When people are run-down or suffering from some kind of infection they often develop blood-shot, jaundiced or otherwise discoloured eyes, and this occurs regardless of race.

But the eyes have more significance than this. Tests with schematic eye-spots show that human viewers of all ages are more responsive to a pair of circles with a dot in the middle than to configurations of either one or three such circles. Apparently, we have an instinctual interest in eyes. Moreover, when the dot in the centre of the circle is increased so as to more nearly fill the circle, the interest of the human viewer is also increased. This no doubt relates to the fact that babies have very large eyes (and large pupils) relative to the size of their head. It is clearly important for babies to appeal visually to adults as much as possible in order to evoke protective instincts in them, so the large eyes of the baby have probably evolved so as to make them more lovable.

Women also have large eyes compared with men, and this has nothing to do with visual acuity. They too seem to have evolved large eyes so as to increase their attractiveness to men (and perhaps to evoke a parental protection response). Furthermore, if a woman's pupils are enlarged for any reason it enhances her attractiveness to men. This may occur either because of low illumination (as in the moonlight or over a candle-lit dinner) or it may be because the woman is interested in and excited by her partner. Medieval Italian courtesans created the effect artificially by using a drug derived from Deadly Nightshade called Belladonna ('beautiful woman'). The enlarged pupil is probably seen as beautiful in its own right, but of course with the typical male ego at play it is likely to be interpreted as sexual interest and receptivity whatever its actual cause.

In a famous experiment by Hess (1975), two identical photographs of a woman were shown to a group of men, one with the pupils touched up so as to make them a bit larger. Nearly all the men rated the girl with the larger pupils as more attractive even though few were able to put their finger on the actual difference between the two pictures. The fact that these enlarged pupils are perceived by men as a sexual signal is supported by the finding of Simms (1967) that women and male homosexuals do not prefer the girl with enlarged pupils; in fact, they prefer girls with constricted pupils – presumably because they are less threatening sexually. Shelley and McKew (1979) have further confirmed the sexual significance of pupil dilation by showing that a girl with enlarged pupils is preferred only by *adult* males; the effect is hardly observable in preadolescent males. In this experiment also, females, both preadolescent and adult, preferred the girl with small pupils.

Hess also studied the pupillary responses of men while they were looking at girls with small and large pupils. This confirmed their preference for the girl with the enlarged pupils, since the pupillary response of the male viewers was more than twice as strong to the girl with the large pupils. This was again despite the fact that the men when interviewed afterwards said they thought the pictures were identical. None remembered that one photograph had larger pupils than the other. Simms (1967) followed this up by looking at the responses of men and women to pictures of their own sex. Women showed a smaller response to a female picture with large pupils than they did to a woman with small pupils (the opposite of the male pattern), while the men showed virtually no response to a picture of a man regardless of pupil size.

Another group that apparently prefer women with smaller pupils are self-confessed 'Don Juans' – men who seek female conquests but prefer to avoid emotional entanglements with any particular girl. In Hess's research such men showed a reversal of the more usual male pattern, producing greater pupillary dilation to the girl with small pupils than the one with large pupils. This could be interpreted by supposing that these men were using pupillary dilation as a warning signal that the girl might be *too* interested in them and likely to be clinging.

These experiments indicate that men have a special interest in the signals of sexual receptivity emitted by women. Another fascinating study by Hess suggests that women may have a natural maternal interest. When men and women were shown photographs of babies, the women showed pupillary dilation regardless of whether they were single or married, childless or mothers. When the baby pictures were shown to men, the result was usually pupillary constriction, the one exception being the case of married men who had children. Apparently, men have to learn to be interested in children after having children of their own. Women, however, seem to be genetically primed for maternal reactions.

Visual arousability

Earlier in the chapter it was suggested that there is a

difference between men and women in the readiness with which they may be aroused by the sight of each other. In all societies, boys and men spend a great deal of time looking at girls and commenting on their physical attributes. Women, by contrast, are very little interested in the visual attributes of men. They may be impressed by a man's skill and prowess in work, sport or battle, but they are seldom excited by the mere sight of a hairy chest and bulging muscles. The difference is even more striking where the sight of genitalia is concerned. Boys and men make great efforts to look at female genitalia, and are fascinated by their infinite variety. Women are aware of this lecherous male interest and take care not to expose themselves in a manner that might provoke unwanted attention. But there is also a sense in which they have difficulty in comprehending this voyeuristic interest of the male since they are not themselves sexually excited by the sight of male genitalia. According to circumstances, they may regard the male apparatus as funny or threatening, but immediate sexual excitement is unlikely to be their prime reaction. As Symons (1979) points out, women can work all night in massage parlours and 'wank shops' without themselves getting aroused. In the reverse situation, this would be quite impossible for men. Even homosexual men would probably get aroused if they were employed to masturbate women.

The most obvious manifestation of this male visual interest is seen in the media and the pornography market. The production of 'girlie' magazines as well as harder core pornography is a multi-million pound industry in Western countries, and it is almost entirely directed at men. Not even lesbians are very much interested in looking at pictures of nude girls. The only magazines which feature nude males to any extent are quite explicitly directed at the homosexual male market. With the advent of women's liberation two women's magazines did experiment with the inclusion of male nudes but the experiment was an almost total flop. There was no interference from the legal authorities, but rather a general lack of interest and support from the female readership. *Viva* reversed its policy of incorporating male nudes shortly afterwards, apparently because the readers found them either ridiculous or distasteful. *Playgirl* barely survives with a combination of women's interest articles and male nudes, and is therefore the only magazine in the Western world which is supposedly aimed at women and which includes nude male pictures. However, there is a serious question concerning the motives for purchasing it since a high proportion of buyers are male. Some of these men may be homosexual; others may be buying it for their girl-friend as a jokey present or an attempt to titillate her. Some of the women buyers probably support it for political reasons as much as enjoying it sexually. In terms of sales success, the female equivalent of 'men's magazines' seems to be either the Mills and Boon romance novel, or the domesticated magazine which features tips about homes, gardens, marriage, children, make-up, clothing and other traditional feminine interests. If these publications portray nudity at all it is more likely to be female nudity than male.

Since the days when Kinsey reported women as being less readily aroused by pornography than males a number of laboratory experiments have been conducted which appear to show that women do respond to pornography and perhaps just as much as men (for example Schmidt and Sigusch, 1973; Heiman, 1975; Gillan and Frith, 1979). These studies typically compare verbal reports of arousal, and physiological indices such as penis volume and vaginal lubrication, to the showing of erotic slides, films and tape recordings. Results of these studies show that women are capable of responding to pornography under laboratory conditions and some people have therefore concluded that the lack of interest shown in pornography by women in the real world must be due to some kind of social conditioning. At first sight this seems a reasonable conclusion, but given a direct conflict between laboratory reactions and actual social behaviour it would be just as well to question the validity of the experimental studies. In fact, there are a number of reasons for thinking that the experiments cannot be held to have established identity between men and women in response to pornography.

The first reason for suspecting the findings is that male volunteers for experiments on reactions to pornography are easier to come by than female volunteers. This means that the women studied are a less representative sample of their gender than are the men. It is highly probable that the direction of bias is towards liberal and women of high libido rather than the puritanical and undersexed; in other words, women who are more like men in sexual attitudes and tendencies. This sample bias would, then, tend to diminish the apparent difference between men and women in response to pornography.

Another major problem with the pornography experiments concerns the comparability of the measurement techniques. Subjective ratings of sexual arousal to pictures and films are of dubious validity because members of the two sexes cannot possibly know how much arousal the other experiences. Probably a man rates his degree of arousal primarily in relation to his own lows and peaks, and perhaps to some extent relative to the way he thinks his male friends would respond. Assumptions about how women typically respond are unlikely to figure prominently in the frame of reference he uses for judging his own arousal. Women presumably respond similarly; their scale for judging sexual arousal would be derived from the range of their own experience, and to a lesser extent that of their female friends. In other words, we cannot be sure that men and women are using the same standards for rating their sexual arousal when exposed to pornography.

We might hope that the physiological measures would be less subjective, and therefore more valid, but in fact they present an even greater problem of comparability. For example, what percentage of increase in penis volume should be considered equivalent to a given increase in vaginal lubrication? Since the techniques used for measuring sexual arousal are different in the two sexes (of necessity because of the anatomical differences) it is virtually impossible to say whether men or women are more 'turned on' by particular stimuli.

One measure of physiological response that is the same for men and women is adrenalin secretion, and this measure shows men to be rather more reactive to pornographic films than women. Levi (1968) showed explicit film clips of sexual intercourse to 53 female and 50 male students taking samples of their urine before, during and after the film showing. Although adrenalin output was higher for both men and women during the film, the change occurring in the male subjects was considerably greater than that registered by the females. Although adrenalin secretion could be reflecting any kind of emotional stress such as fear or anger, it is also known to increase with sexual excitement, and the verbal reports of the subjects confirm that this was their primary emotional experience.

Another unisex measure of arousal, pupillary dilation, shows that males and females react about equally to pictures of the opposite sex (Hess, 1975). At first sight, this might seem to contradict the findings concerning adrenalin secretion. However, pupil dilation is really a measure of interest rather than arousal, and the women might be interested in men without being sexually aroused by them.

Although the laboratory studies do show that women are capable of being aroused by pornography, many of the studies show differences in the *kind* of pornography that is most arousing to each sex. Men are usually found to be more responsive to visual pornography while women are relatively more stimulated by auditory and literary modes of presentation (Byrne and Lamberth, 1971; Gillan and Frith, 1979). The mere sight of male genitals is not arousing to women in the same way that female nudity is to men. Women are really only turned on to pornography that includes a sequence of activity between a man and a woman, preferably one that develops slowly and progressively. A number of writers have suggested that what is important is that the woman is able to use her imagination and project herself into the sequence, perhaps by identifying with the woman in the story (Jakobovits, 1965; Symons 1979). Steele and Walker (1976) studied the responses of female students to a variety of films and interviewed them to probe specific likes and dislikes. On this basis they arrived at a description of the ideal erotic film from a female perspective. 'The cast of the film would consist of one attractive male and one attractive female displaying affection, "romance" and prolonged foreplay in a bedroom setting. The film would involve a *gradual* process leading to coitus involving a variety of positions. The emotional tone of the film would emphasise the "total" relationship, and not merely genital sexual behaviour.' It seems that while women may be aroused by films almost as much as men, the mechanism of arousal is different and so, therefore, is the optimal style of pornography.

If there has been a change in female responsiveness

since the time of Kinsey it may well reflect the quality and complexity of the material that is now available. In Kinsey's time such pornography as was available was nearly all sordid and male-oriented. Today, with a more permissive social climate, we are seeing better produced, more subtle and yet active pornography (perhaps better called 'erotica') and this is a great deal more unisex in its appeal. The *Emmanuelle* series starring Sylvia Kristel is a good example of film material that incorporates elements of both male and female fantasy. *Forum* magazine is also a deliberate attempt to appeal to a unisex market; there is plenty of open discussion about physical sex, but much of it within the context of loving, personal relationships.

In any case, regardless of the experimental evidence, it is quite clear that in the real world women do not seek out visual stimulation to the same extent that men do. There are similar sex differences in the content of the graffiti that appear on toilet walls. Girls produce mainly verbal, romantic graffiti – non-sexual statements of attachment – while boys are much more given to pictorial, explicitly sexual graffiti (Wales and Brewer, 1976).

The visual emphasis of men is also seen in their sexual fantasies. Barclay (1973) had 150 college men and 150 college women write full descriptions of their favourite sexual fantasies and these were analysed for sex differences. One of the most striking characteristics of the male fantasies was their resemblance to features from pornographic books. There was a strong emphasis on voyeuristic aspects, with many men being able to report minute details of their partner's physical appearance and the activities they engaged in. The female characters appearing in men's fantasies tended to be seductive and willing and there was a general absence of personal or romantic involvement with them. By contrast, the women reported fantasies in which the man was known or in some way identifiable (for instance a neighbour or film star). At most, a description of hairstyle or eye-colour was obtained; never any details about his body or genitals, or the mechanics of intercourse. Women were generally more concerned with the quality of the sexual experience and their degree of romantic involvement with their partner. Such visual images as women did report more often referred to the setting (such as visions of the sun, beaches, green fields and flowers), than to the physical attributes of the partner.

These differences between men and women in the ease with which they may be visually aroused have frequently been interpreted as reflections of social learning processes – men learning that it is socially acceptable (or even expected) to be 'turned on' by the sight of a woman, and women learning to be demure and to control their arousal. Yet the difference seems too stable and universal to have arisen solely out of social learning, and an evolutionary explanation is readily apparent. The male reproductive strategy of having many casual sexual encounters with a wide variety of women is well served by ready visual arousability. Biologically speaking, it is to their advantage to get aroused by every woman they see (within reason) and to attempt to mate with them. Such a reaction in women, however, would be counterproductive in terms of their optimal strategy of being highly selective and requiring demonstrations of long-term interest and devotion from potential suitors. If women were to become sexually aroused by every man they saw, they would have great difficulty in sustaining such a strategy. Thus it appears that males have evolved as sexual hunters; they are visually attracted to females from a distance and this is enough to evoke sexual interest and initiate advances. Women, on the other hand, have evolved mechanisms for avoiding being turned on by the mere sight of the opposite sex; normally they require special attention from the male, some evidence of his fitness (usually skill and prowess) and often some degree of tactile stimulation before they become sexually aroused.

Choice of partner

In many cultures one is not permitted to choose one's own partner; marriages are arranged by older members of the family or tribe. This has sometimes been thought to mean that we do not naturally form sexual preferences among different people – that one person of the opposite sex will do just about as well as another. On the contrary, people do have strong tendencies to form sexual relationships of their own, but the family have their own motives for overriding these preferences. Usually, they do so on the argument that the individual is blinded by temporary passion and that their own choice is based upon criteria that have enduring importance. But even in societies that arrange marriages, it is not unusual for the young people to resist the will of the elders by elopement. A highly publicized recent case was that of the Saudi Arabian Princess who was apparently forced into a marriage that was not to her liking and was subsequently executed for running away with a student that she loved.

Variations in beauty and social status are of prime importance in determining the choice of a partner. Also, as we have seen in previous chapters, novelty is a powerful basis of preference to the male animal of most species, including humans to some extent. Another fairly obvious factor determining female attractiveness, especially in lower animals, is the state of their cycle; rats, for example, are really only attractive to the males in oestrus. But even when such factors are taken account of animals show striking individual preferences.

Beach (1975) has described these consistent, though

seemingly irrational preferences amongst different available partners in dogs.

> One of the females was named Peggy. She had grown up with and played with all five males. When she came into heat they all wanted to mate with her. She was very happy to mate with the male named Broadus, and she was willing to mate with three others, but she simply would not mate with the fifth male, Ken. She knew him well; they weren't enemies. As a matter of fact they got along quite congenially when she was not in heat, and he clearly was socially dominant over her, but she would not mate with him. When he persisted in trying, she attacked and bit him until he bled. I followed this particular pair through six years. Peggy never lost her antipathy for Ken as a sexual partner and she never lost her positive attitude toward the other males.
>
> Studying the behaviour of other bitches, I found that most if not all of them have definite sexual preferences. They are much readier to mate with some males than with others. In extreme cases, a female simply won't receive a given male even though the same male may be acceptable to a different bitch. One of the interesting things was the consistency of this behaviour; the patterns of likes and dislikes persisted right through the years. Even when we brought the female into heat artificially by injecting ovarian hormones, it was the same story. In a couple of cases I gave double or triple the normal hormone dose to see if I could override the preference, but I couldn't.

Beach was never able to work out the basis of these striking sexual preferences, but they have been observed in most other species – for example rats (Drewett, 1973) and gorillas (Nadler, 1975) – and they are usually found to be more decisive in females than males. Males have their preferences alright, but they are seldom totally averse to having sex with females low on their preference list. Within the male strategy it is 'any port in a storm'. For some men any woman is better than masturbation. Anthropologists have noted that in many societies certain girls are almost universally described as so ugly that no man would consider having sex with them; yet these girls always seem to manage to get pregnant, if not necessarily married (Symons, 1979). Females, on the other hand, both human and animal, develop compelling attractions to certain males and also some very definite exclusions. Quite literally, many women would not sleep with certain individuals 'if they were the last man on Earth'. This policy is of course consistent with the female strategy of taking care not to squander their limited breeding potential.

Another difference between men and women concerns the extent of agreement within the gender about which members of the opposite sex are attractive and unattractive. If men are asked to rate the desirability of a selection of girls, there will be a high level of agreement amongst them, whereas women have relatively idiosyncratic criteria for assessing men (Wilson and Nias, 1976). This difference could reflect the fact that men place a great deal more emphasis on physical looks, while women are more concerned with intellectual and social criteria. However, given the female predisposition to form strong and exclusive attachments to particular men, it is perhaps just as well that nature has arranged it that all women are not simultaneously pursuing the same man. To a woman, it is much more important that a man be eligible and attentive to her than physically desirable.

Avoiding incest

Nearly every society has strict rules forbidding sex between close family members. While there may be various attitudes towards sex between lesser-degree relatives such as cousins, sexual relations between father and daughter, mother and son, and brother and sister (at least in adulthood) are universally forbidden. Often they are regarded with great repugnance and transgression is very severely punished.

The most widely held theories of the incest taboo are framed in psychoanalytic and sociological terms. Freud and his followers argued that such a powerful prohibition, backed as it is with religious and legal sanctions, would not be necessary unless the urge to engage in incestuous behaviour were widespread and strong. Thus Freud formulated his theory of the Oedipus and Electra complexes, according to which little boys and little girls around the age of four and five are supposed to fall in love with their parents of the opposite sex and to desire them sexually. They soon learn, however, that such lust is unacceptable and dangerous, and so it is repressed (relegated to the unconscious mind) throughout the rest of their lives, perhaps manifesting itself indirectly in the individual's choice of sex objects and marriage partners. The feeling of disgust that most of us register at the idea of having sex with a member of our immediate family is supposed by the psychoanalysts to be a kind of 'reaction formation' by which we prevent ourselves from becoming aware of our deep-seated but unacceptable desires.

The most popular sociological theory of the incest taboo also supposes that we have a natural desire to have sex with other members of our family. However, the sociologists maintain that we have developed strict rules to restrain ourselves from submitting to these desires, so that family cohesion and roles are not threatened. Sexual relationships within the family, they say, could create rivalries, jealousies, and general dissension that could easily destroy its effectiveness as a friendly and supportive social unit. Most obviously, parent-child liaisons would threaten the marriage, but siblings could also be jealous of each other and their parents.

Neither of these theories fits very well with the facts of incest avoidance. The actual occurrence of incest, other than prepubescent playfulness between brother and sister, is so rare as to make it highly improbable that we have any real urge to indulge in it. If our desire were really so burning as theorists have supposed we would surely find the outlet for it occasionally, whatever the strength of the sanctions against it. Even if the urge were 'unconscious' in the psychoanalytic sense, we would expect it to break through in our dreams and fantasies every so often. However, our own research (Wilson, 1978; Gosselin and Wilson, 1980) indicates that incest fantasies are virtually non-existent in both normal and sexually variant groups. The Freudian and sociologial theories would also have some difficulty in explaining why monkeys and chimpanzees avoid incest as well as human beings (Demorest, 1977). After all, non-human primates should be relatively free of social inhibitions of the kind implied by taboos and legal sanctions. These theories would also have difficulty in explaining why mother-son incest is so much less common than the father-daughter variety.

Altogether, there is insufficient reason to suppose that the average person has any real positive attraction to incestuous sexual contact, either consciously or unconsciously. It is much more likely that the taboo is a social expression of a concomitant instinctual preference – an effect of our feelings rather than a cause of them. Perhaps the best comparison is with the crime of murder, which is even more taboo and repugnant in most societies than incest. Our strong aversion to the thought of murder does not mask any desire in the normal person to go about liquidating his fellow man. In both animal and human societies murder is a rare event which occurs only under exceptional circumstances, and the strength of the legal sanctions against it have very little effect on its frequency. Similarly, when incest does occur, it is usually possible to identify exceptional circumstances which account for it. Father-daughter incest is most likely to occur in remote, rural families in which the wife is for one reason or another incapable of fulfilling her role (due, for instance, to death or disability) and the father can find no alternative partners because of geographical and social isolation (Bagley, 1969). Mother-son incest is extremely rare, and when it does occur one or both of the participants are usually found to be mentally retarded or psychotic (Lukianowicz, 1972).

Even in the two most famous classical accounts of incest we can identify extreme circumstances. Lot and his daughters (*Genesis* 19) are the sole survivors of the cataclysmic destruction of Sodom and Gomorrah, the sons-in-law having stayed behind for the last orgy, and the wife having been turned into a pillar of salt for hesitating in the flight. Therefore the daughters felt justified in seducing their father in order to continue the family line and recreate a new, purer civilization. Even so, they were obliged to get him so drunk that he did not know what (and especially who) he was doing. Interestingly, this Biblical account of incest may be partly responsible for the fact that father-daughter incest is most likely to occur in fundamentalist religious families, the father taking this text as his example and believing that no man is virtuous enough to have sex with his daughters other than himself. In the Greek legend upon which Freud based his theory of the Oedipus complex, Oedipus is not even aware of the fact that the queen he has married is his own mother; nor is she aware that Oedipus is her son. When they find out their true relationship, she commits suicide and Oedipus gouges his eyes out with her brooch. There is nothing in either of these stories to suggest that the actors were motivated by unsocialized lust. They are either unaware or respond to force of circumstances. It is also worth noting that neither myth has much value as erotica.

If we allow that the facts of incest are more in accordance

with the idea that we have an instinctual preference for having sex outside the family, it is easy to provide an evolutionary explanation. Because close family members are likely to carry the same recessive deficiencies on their genes, there is a greatly increased risk that these will be manifested in their offspring. This is true not just of the relatively common and harmless genetic anomalies such as colour blindness, but also of more extreme disorders that give rise to mental retardation, serious physical debility and death. The disadvantageous effects of inbreeding have been observed clearly in certain parts of rural Ireland and Tennessee, while enhanced intelligence in the offspring of mixed-race marriages has been documented in Hawaii. There is little doubt that we give our genes a better chance of survival by mixing them with an unrelated set.

There are two commonly raised objections to this theory, neither of which has any substance. One of them is based on the idea that human societies would have to *recognise* the deleterious effects of inbreeding before being motivated to avoid incest (Katchadourian and Lunde, 1975). This is of course a misconception of sociobiological mechanisms. Genes store a great deal of information serving their survival that does not require a full logical understanding of the workings on the part of the host animal before it can be implemented. When an animal copulates it has little awareness of the procreative significance of the act; nevertheless it feels compelled to do so. Many human societies were also late in discovering the connection between intercourse and pregnancy. Clearly, consciousness of the benefits of avoiding incest is not a necessary condition for its effective operation.

The other objection is that, by emphasizing whatever characteristics are available in the family, inbreeding might just as well be beneficial as detrimental. The case that is most frequently cited is that of the Darwins, Huxleys, Galtons and Wedgwoods, who were closely interrelated families that produced a great many brilliant scientific and literary men. Sir Francis Galton himself, somewhat immodestly, described this as an example of inbred genius. It is, in fact, possible to produce superior stock from the inbreeding of high quality families, which might explain why some societies permit or even encourage their ruling families to breach the incest rule. The Azande of Africa, for example, encourage the highest chiefs to have sex with their own daughters. Intermarriage between siblings was permitted for Inca chiefs and the Egyptian pharaohs. Cleopatra was both born of such a union and herself a participant in one. In more recent times, the European royal families have shown a degree of inbreeding, although not to an extent that could be called incestuous. Even so, haemophilia appeared in the British royal family in the last century, as a reminder that harmful recessives can combine to assert themselves in the offspring of related parents. While inbreeding may sometimes produce beneficial effects, the danger of disease and destruction is so much greater that most geneticists agree that avoiding incest does have significant survival value.

Evolutionary theory can also account for the fact that father-daughter incest is very much more common than mother-son incest. We have argued that males have a kind of in-built tendency towards promiscuity because it is to their reproductive advantage to fertilize as many females as possible. This drive could be extended to include a man's own daughter, even though she is likely to be low on his list of priorities as a sex target for the reasons given above. The woman's optimal strategy is quite definitely to avoid incest because of the chance that she might waste her limited opportunity for reproduction in producing offspring that are imperfect and less able to survive. It follows that mother-son incest is extremely rare and associated with gross psychiatric disturbance. This analysis treats the parent as responsible for the occurrence of the incest rather than the child, regardless of who appears to have instigated it. This is probably fair, for clinical observation suggests that the children are usually acquiescent in such relationships.

The greater frequency of father-daughter incest compared with mother-son incest is also consistent with the usual age differential in mating discussed above. It makes a great deal more biological sense for an older man to be associated with a younger woman than the other way about.

The harmful effects of inbreeding may explain why tendencies to avoid incest have evolved, but they do not explain the psychological mechanism by which the functional behaviour is achieved. Since no evidence has been adduced that we have power to identify and reject close family members we have not grown up with, the most likely possibility is that constant exposure to other family members dulls the potential for sexual excitement ('familiarity breeds contempt'). This theory is supported by the observation that children who grow up together in close proximity seldom get excited by each other in adolescence whether they are blood relatives or not. The 'new girl' or 'new boy' in town is much more likely to arouse local passions. Similarly, non-human primates reared together from infancy have been observed to show a kind of incest avoidance (Demorest, 1977). The tendency for married couples who have lived together for many years to become sexually bored with each other could also be interpreted in such terms.

There is no doubt some truth in this proposition; the kinds of interaction that occur between members of a family unit (attempts to control and manipulate each other, fights, jealousies, etc.) are bound to be antithetical to the development of romantic and sexual interest. Still it is unlikely to be a complete explanation. The learning of social norms and taboos probably reinforces these naturally developing asexual attitudes among family members. It is also fair to concede that the social functions of avoiding incest (safeguarding the integrity of the family and building larger social units through kinship ties) could also contribute to our survival and therefore add to the evolutionary advantage of such a behaviour pattern. Still, there is no need to suppose that the strong taboos against incest have been put up by 'society' in opposition to the individual's sexual instincts. These two forces, the personal and the cultural, are for the most part working in concert.

5

Love and Marriage

The experience of 'true' love is a central theme in the popular media. Books, magazines, newspapers, films and television devote a great deal of attention to it and there can be no doubt that it is a major preoccupation with most of us. Love is portrayed as entrancement with another person, normally coupled with a desire for sexual contact with them. There is a widespread belief that we are really only happy and fulfilled when we are in love with someone, and we may be persuaded, like Gilbert and Sullivan's Patience, that whole-hearted, self-sacrificing love is nothing less than a social duty. However, it is also recognized that love may be a distressing and anti-social force. Sometimes Cupid's dart strikes irrationally and inconveniently so that the individual is seen by society (or even himself) as victim of a kind of disease (Pigache, 1978).

The relationship of romantic love to marriage is also ambiguous. In our society, love and marriage are said to 'go together like a horse and carriage'. Love is regarded as a necessary, if not sufficient, cause of marriage. But there is also a conventional wisdom to the effect that love has difficulty in surviving the formal state of marriage for any length of time. As Oscar Wilde put it, 'Love is a temporary insanity curable only by marriage.' After marriage, the passionate craving for the partner is likely to wane progressively, and at some point true love may strike again in another direction. How this reappearance of passionate love is handled depends upon the culture. In California, one usually gets divorced and remarried to the new lover. In France, one remains married and pursues a more or less discreet affair with the other person. In cultures that arrange marriages, love is not seen as a prerequisite for marriage at all but rather as a separate phenomenon; when it does occur it may be tolerated as perfectly natural or punished severely.

There has always been resistance to the idea of studying love from a scientific point of view (Wilson & Nias, 1976). Of all our emotions it is viewed as the most ecstatic and sacrosanct, the most complex and mysterious. Many people have supposed that it is uniquely human, a reflection of our proximity to God, and therefore not amenable to biological explanation. But this is again a species-centred conceit that is not supported by the facts. Many animals form close-knit male-female partnerships that are indistinguishable from our own in objective terms, and certain 'lowly' birds are more faithful to their partners and more likely to die of a broken heart than we who regard ourselves as next to the angels.

In discussing the evolutionary and instinctual bases of love and marriage, there are two fairly separate questions that need to be asked. The first is why romantic love and monogamy occur at all. Why do we fall in love and establish more or less stable partnerships? The second is why we fall in love with a particular person rather than someone else. How do we come to select our love object? We shall deal first with the general question concerning the occurrence of monogamy and marriage.

Conditions favouring monogamy

We have noted that there are two types of polygamy in the animal world. One is *polygyny*, which involves one male associated with a harem of females; the other is *polyandry*, in which one female enjoys the attention of several males. *Promiscuity* is distinguished from both of these in that it implies fairly random matings. Of these arrangements polygyny is by far the most common, at least in the mammalian world, and this is exactly what we would expect on the basis of parental investment theory. In those rare species which engage in polyandry, there is usually also a reversal of parental investment, with the male contributing a great deal of time and effort to raising the offspring (Wilson, 1975).

Monogamy, the pairing of one male with one female at least for the duration of one mating season, also occurs under some conditions, although more often in birds than mammals. In particular, monogamy seems to evolve when environmental conditions are so difficult and hazardous that the offspring may not survive without the support and protection of both parents. This means that a species will tend towards monogamy when territory and food are in very short supply and when there is a risk of heavy predation from other species. Otherwise, when there is an abundance of space and food, and when there is no great risk of enemies preying on the young, polygyny tends to emerge as the natural order. From the female point of view it may be thought of in this way: 'If the environments of different territories vary sufficiently in quality, a female will gain more in genetic fitness by joining other females in the single rich territory of a polygynous male, than by becoming the sole partner of a monogamous male on poor land' (Wilson, 1975).

How does this apply to the human condition? Generally

it is true that rich men manage to gather around them a greater number of women than poor men. In some cultures they may become official wives and live in a harem; in European society only one at a time may be legally married even to a rich man, and the rest are called mistresses or girl friends. Clearly, polygyny is quite common in human society, and the men who control the greatest number of women are those with prime real estate and ability to provide the most food. In this respect, man takes his natural place among most other mammals and primates.

There are, however, two forces unique to humans that seem to push us back towards monogamy to a greater or lesser extent. One is the sense of morality (fairness or justice) that we have developed along with the evolution of our exceptional brains. Actually, the evolution of altruism, with its power to promote peace and cooperation, can also be explained in terms of its effect on group survival, and thus perhaps reduced immediately to an amoral concept. Nevertheless, it does exist as a particularly human instinct, and it has led some societies at least to promote the view of 'fair shares', not just of property, but of women as well. The most notable institution propounding this view in our society is the Church, although the principles of communism might also be read as implying monogamous marital arrangements. Hence a respectable family life may be an essential qualification for many positions in public life. Many people believe that Nelson Rockefeller could have become President of the United States were it not for the fact that he had been divorced, and Teddy Kennedy's recent campaign for the Democratic nomination was certainly not enhanced by his domestic instability. Such moral considerations appear unique to the human species and may acount for a greater degree of monogamy than would otherwise be the case;

although, who knows if a monkey was ever given credit in monkey society for being satisfied with a single mate?

The second development within human evolution that may have led us half-way back towards a bird-like pair bonding system has been suggested by Desmond Morris (1971). Morris maintains that humans have become a great deal more obsessed with eroticism than other apes, not because of the media but because of our biological evolution. He notes that as a species we have lost nearly all of our hair (especially women) and since this is otherwise disadvantageous it must be because of the erotic value of bare flesh. He further points out that our females are receptive all through the month, not just during ovulation, and their breasts are permanently protuberant, not just when suckling babies. All of these things, he says, make humans the sexiest animals in creation. Add to this the importance of individual identification that is tied up with our face-to-face habits of communication, and we have the conditions for developing strong intimate bonds with particular individuals that are constantly reinforced by sexual pleasure with that same person. Presumably the ultimate advantage of this is that stable social living is possible and lethal jealousies are held in relative check.

Not all biologists accept Morris's argument. Another possible reason for the human loss of hair is that nakedness allowed superior sweat-cooling during the strenuous pursuit of prey across the African savannah. Certainly the human body is much better equipped with sweat glands than any other primate species. But Morris would probably want to know why, in that case, men are hairier than women; if anything it ought to be the other way about. One other suggestion is that our smooth skin allows us to swim faster than other apes, which might be appropriate to a shore-living primate with a taste for sea-food. Again, this explanation is rather unlikely considering that the early hunter-gatherers were probably men, who are hairier than women.

Marriage and the loss of oestrus

But humans are not really pair-bonding animals in the sense that parrots, penguins and ducks are. In terms of our natural inclinations we are, just as would be expected, rather like other apes – basically polygynous harem-builders with a capacity for forming strong friendships and short-term exclusive sexual liaisons. Some critics of this position point to the gibbon as an alternative model for human mating behaviour since this is a primate that does seem to form social groups comprising a pair-bonded male and female with their own offspring who defend a well-defined territorial area against intruders of either sex. Superficially, this would seem to resemble the human nuclear family. But the analogy is unsatisfactory for two reasons: (1) Gibbons are not so closely related to us as the great apes. They differ in the important respect that the males and females are not differentiated in physical size and, as pointed out in the previous chapter, the more the male exceeds the female in size the more polygynous the species tends to be. (2) While gibbon pair-bonding is fairly spontaneous and independent of the social context, human marriage is very largely imposed by social pressures for social purposes.

> . . . in most important respects the gibbon family and the human family are almost exact opposites; the gibbon family exists because adult males and females repulse from their territory same-sex conspecifics, and the family is both the smallest and the largest gibbon social group: the human family, on the other hand, does not really exist apart from the larger social matrix that defines, creates and maintains it. For the greastest majority of humanity – and possibly for all

> of it before modern times – marriage is not so much an alliance of two people but rather an alliance of families and larger networks of people. Among most non-modern peoples marriages are negotiated and arranged by elders, not by the principals, although the latter may have a say in the matter; in some cases a girl is betrothed before she is born. Marriage begins with a public pronouncement – and usually a ceremony – and can be said to exist only insofar as it is recognized by the community at large. Obligations and rights entailed by marriage vary among societies, but marriage is fundamentally a political, economic, and child-raising institution, based on a division of labour by sex and economic co-operation between the spouses and among larger networks of kin.

Symons goes on to note that insofar as strong emotions are involved in human marriage, jealousy (especially male jealousy) is most significant. The strong positive sexual interest that may be involved in the initiation of the marriage is usually dissipated within a short time afterwards.

Thus does Symons argue against the Morris idea that human beings are natural pair-bonders and that marriage is the social manifestation of that pair-bonding. The human marital relationship is not founded on sexual passion but on practical considerations such as hard work, loyalty, even temperament and adherence to social custom. Sexual attractiveness may enter into the initial choice (particularly when a man chooses a woman in Western society) but it is unlikely to remain the basis of the marital tie. The so-called pair-bond more often survives because of social pressures, 'duty' and concern for the children, than because of perpetual lust for, and enchantment with, the spouse. After a certain period of marriage most people's libido, sexual fantasies, and 'romantic love' are likely to be directed elsewhere.

The fact that marriage usually outlasts the love-bond is

highlighted in some research by Judy Todd and Ariella Friedman of California State University, Dominguez Hills (1980). They reasoned that marital counselling practice would benefit from an examination of the characteristics of couples whose marriages had lasted for fifty years or more without breaking down. The presumption was that if they were still together they must have been very happy with each other. Unfortunately, nothing could have been further from the truth. Most of them reported having shared fifty years of barely mitigated misery. 'We would get so sad at the end of an interview that we wouldn't do another one for two days,' said Todd in a *Los Angeles Times* interview. 'Then we would go up to the door of the next one, our tape-recorders in hand, praying for a happy one this time.' What the researchers learned was that for these people the marriage, not happiness was important. Their attitude was one of pride that they had survived the depression, they had survived the war, and they had survived their marriage. 'Here we are, two souls surviving in a world of distress. We made it.' That they had lived together for fifty years was not a reflection of happiness so much as duty or social conditioning; they did not regard divorce as an option.

There is little evidence for the idea that repeated copulation (made easier by women's loss of a tightly defined period of oestrus) is a major reinforcer of either romantic love or marriage. In fact, the impression gained from history and literature is that romantic excitement can be stimulated as much by sexual frustration as by sexual fulfillment. A separation between the erotic and affectionate components of sexuality is also indicated by hormonal studies (Rossi, 1973). Women who have had their adrenal cortex and ovaries removed surgically, thus being deprived of all androgens, show a marked drop in libido but no loss of capacity for affection. Human love apparently does not depend upon continuing sexual activity.

All of this suggests that the extended period of sexual receptivity in the human female has not evolved in order to cement the marital pair-bond. In fact, marriage appears to have very little to do with sexual attraction. It was suggested in the introduction to this book that the loss of oestrus was probably due to the increased control that women thus gain over the deployment of their sexual favours. If they are not indiscriminately receptive at the point of ovulation, then they are in a position to make better choices among the males who are offering services, selecting those of better breeding stock. They are also better placed to choose a time when the environment is favourable to the upbringing of children.

Both these functions can now be carried out by contraception. Sex can be enjoyed as recreation with the help of contraception, and it can be used for sensible procreation by withholding it. But again, it must be remembered that our sexual instincts evolved well before the advent of modern birth-control, so this is not a factor to consider in examining the biological basis of human sexuality.

There is another way to view the change in the cyclic nature of female sexuality from ape to human. Apart from the smoothing of female libido through the course of the menstrual cycle, there is also an extension of attractiveness such that men find women sexy all the time – not just when they are ovulating. There is a sense in which the human female is masquerading as fertile all of the time, regardless of whether or not she is ovulating. The swellings of her bottom, 'echoed' in her breasts, are a constant sexual stimulus to the viewing male, so it is reasonable to ask what advantage this may be to the human female.

Symons (1979) has produced a plausible theory as to how this continuous attractiveness of the human female may have evolved. He notes that among chimpanzees the hunter males are more generous in sharing their catch with oestrous females. As hunting became more and more a male activity during human evolution it became progressively important for the females to persuade males to share meat with them, and faking oestral signals for an extended period helped them in doing so.

> Optimal male strategies in distributing meat may depend in part on whether they were able to detect ovulation. If hominid males regularly possessed meat surpluses before oestrus was lost, a good hunter might do best reproductively by exchanging meat for copulations with oestrous females. If ovulation could not be detected, however, a successful male might be better off acquiring permanent sexual rights to a female or females, resulting in relatively high confidence in paternity, male provisioning of his mate's offspring, and the evolution of other kinds of paternal behaviours and predispositions (Symons, 1979).

Thus, in concealing ovulation and permanently advertising her charms, the human female not only became a better prostitute (with a longer open season), she possibly created the incentive for men to marry her (that is, to acquire exclusive sexual rights over her). This explanation, then, is a slight variant on the older ethological argument to the effect that continuous female receptivity evolved to foster social ties. In this view, marriage evolved not because humans had become sexual pair-bonders, but as a form of property right that was socially acknowledged and enforced.

Jealousy and the motives for marriage

If the basis of marriage is ownership rather than sexual attraction, then it is easy to see why jealousy is the prime emotion engendered by the state of marriage rather than love. The emotional strength of male jealousy is well illustrated in the phenomenon of wife-beating. Recent research in the Glasgow area by Dobash and Dobash (1980) shows that serious violence done to wives by husbands is about seventy times as common as equivalent violence directed at husbands by wives, and that men attack women (usually their wives) about as often as they assault other men. While drinking was associated with a certain proportion of wife-beating incidents, the authors were convinced that it was not the fundamental cause. Instead, they calculated that nearly half of all wife-beating is precipitated by sexual jealousy. Other causes such as rows over money, housework or the preparation of meals were much less common. The authors also concluded that a major reason for the continuance of wife-beating was acceptance by all the social agencies concerned that a husband is entitled to discipline his wife for sexual misconduct.

Although husbands often do feel murderous towards an unfaithful wife, it is equally common for society to exact the punishment against an adulterous woman, sometimes even against the wishes of the husband, who has in the meantime developed a companionate love for her. A well-known mythical example that comes to mind is King Arthur who, in the Camelot legend, was compelled by law to have Guinevere burned at the stake for her relationship with Lancelot, even though he had no personal desire to punish her. And in recent times, an English woman in Saudi Arabia might well have been stoned to death for adultery had not her husband pleaded in her defence that he had given permission for her to have sex with the other man.

The treatment of wives as a form of property is also obvious in many preliterate societies, such as the Australian Aboriginals, who will loan or trade wives for political or economic purposes. In our own society, the phrase 'wife-swapping' has been identified as 'sexist' but is still well understood, while the concept of 'husband-swapping' makes little sense to most people. Clearly, human society is more concerned about the property rights implied by marriage than it is about the feelings of the people involved.

Marriage, then, is not necessarily connected with either romantic love or the pair-bond. The instinct that more directly underlies it is that of sequestering a mate and protecting her from other males so that the husband can be sure of his paternity. This is seen to a limited extent in chimpanzees (our nearest animal relatives) but it is rather more important in human society because the father invests considerably in the future of his offspring. If he is tricked into rearing the offspring of another man, the cause of his own genes' survival is certainly not served. On the other hand, if he can successfully defend his wife against intruding males, sequester several wives, or service girl friends in addition, he is undoubtedly operating a better strategy.

Traditional society does seem to be 'male-oriented' in the sense that it endorses male ownership of wives, although it is also in the interest of wives to be loyal and monogamous because the wife that appears to be unfaithful will risk desertion by the husband who can no longer guarantee his biological share in the children.

What of love? Our society (represented in its extreme form in California) probably comes closer than any other to treating love as a basis for marriage. This is a good thing in that the individuals concerned are totally free to choose their partners, and for a while at least, there is no conflict between love and marriage. But the problem is that the human pair-bond is not so rigid as that between birds (or even gibbons) and, despite the popularity of the ideas of ethologists such as Morris, there is little evidence that repeated sexual intercourse consolidates the love bond. Human love (of the passionate and romantic kind) is often short-lived anyway, and it is possible that the strain of living together in relative isolation from the rest of the community hastens the decay of the love bond. Eventually the two parties may start to develop romantic and passionate attachments outside of the marriage. If one of these is defined by the pair concerned as love, separation or divorce is likely to follow (particularly in California) and a new marriage, or perhaps even several new marriages will be established following an 'all-change' reassignment of affection.

By this process of 'serial polygamy', Western society (or at least some parts of it) manages to keep love and marriage quite closely related. When love fades the marriage is disbanded and another contracted with the new love object. This pattern is, however, very unusual on a world scale. Far more commonly, marriages are organized by families for economic, political and social purposes, and individuals fall in love with each other quite independently of marriage.

Some societies mix the affectional and economic bases of marriage in complex ways. In mediaeval and restoration Britain it was not unusual to find what have been called 'three decker' marriages. The first marriage was arranged for a young woman or man to an older spouse for financial reasons. On being widowed, the individual then contracted a second marriage by mutual choice (love). The third marriage was arranged the other way around for a rich widow or widower to a young spouse. Examples of all three tiers can be found in the royal lineage. According to Sir Anthony Wagner in his book *Pedigree and Progress*, both the Wydvilles and Tudors owed their rise largely to royal *mésalliances* corresponding to the first and third tiers of that system.

Desertion and cuckoldry

The theory of parental investment has a great deal to say about the conditions under which one mate or the other will desert the family. Generally speaking the male animal is more likely to desert because his investment in the offspring is less than that of the female. This is especially so at two early junctures (1) just after copulation and (2) at the appearance of pregnancy. At each point the male has greatly increased his chances of passing on his genes, even though there is a slight risk that the female will find the task of rearing the offspring alone overwhelming and will abort the project at some point. This may be done by deserting the infants immediately after birth or perhaps even killing them. Of course, the father might stay around to help ensure that the offspring make it to adulthood, but in the process he is losing valuable time that could be spent inseminating other females. This is the sociobiological dilemma that faces the male animal, and it is one that is seen in many men as well, even though they will not be consciously formulating such amoral equations. In the USA about twenty times as many men as women desert their family before the children have grown up.

A female animal will be more likely to desert her male mate either before the pregnancy takes place at all, or when the offspring are past the stage of needing a mother's care. Naturally enough, she is much more likely to desert when she is still young enough to conceive more children. Males are likely to desert right at the outset or at a point where the children are fully independent and the female mate is beginning to pass the age of practical infant rearing. Instincts of this kind probably still exist in humans, but they are to some extent suppressed by social prohibitions, for example the moral pressure that would be exerted to prevent a husband leaving a newly pregnant wife. The alternative, adopted by men more commonly than women, is that of having a lover on the side. Again, there may be social pressures against such behaviour but they have very powerful instincts to combat.

Males have an instinctual fear of being tricked into investing in offspring that are not their own. While motherhood is a fairly public and unequivocal event, it is often difficult for a man to be sure that he is the real father of his children. This is one reason why males have a tendency to be more upset by sexual infidelity than females. When women are jealous it is usually because they fear that a rival will steal away the time and economic resources of their husband, not an ounce or two of his semen. But men frequently become insane with jealousy at the mere thought of their woman being touched by another man, and no doubt this is because, at the instinctual level of calculation (which of course is emotional rather than cognitive), they are afraid of being caught in the situation of accepting parental responsibility for a child that does not carry their genes.

It is interesting to note that this fear of cuckoldry and the harshness of the reaction against adultery is greater in Mediterranean countries where birth control is not widely used than in the North European countries where the widespread use of birth control has enabled some separation of sex for recreation from sex for procreation. This would seem to imply some degree of conscious control of sexual jealousy. On the other hand, the fear of cuckoldry cannot be an entirely logical reaction against fear of raising other people's offspring, because many men find it comfortably within their hearts to adopt children who are not their own biologically, or have their wife artificially inseminated by sperm donated to a hospital bank. It seems then, that the jealousy instinct operates at the point of not being able to tolerate another man penetrating one's wife.

Husbands have much less difficulty allowing their wife to be artificially inseminated when the donor is not himself present and participating on the job. This is not surprising when it is considered that artificial insemination is a new technology of which the evolution of instincts could not possibly have taken account.

Relevant to all this is the apparent greater capacity of women to tolerate each other's presence in a group sex situation than it is for men to tolerate each other. Couples who are experienced in such diversions usually report that the presence of an extra woman is much more acceptable all round than that of an extra man. This could be partly because women, being more kind and nurturant by nature, find it easier to relate to each other than heterosexual men do. But mainly it is because the natural orientation of one man towards another is competitive and aggressive, and especially so when the favours of a female are at stake. For while several women can easily be inseminated by one man, and the woman who is the established mate of that man may feel no great insecurity, it is most rare for two men to be able to simultaneously impregnate one woman. (It does happen very occasionally, producing non-identical twins with different fathers.) Therefore the primary male companion of the female runs the risk that she will be impregnated by the other man instead of by himself. Once again, we should not think of these considerations as being conscious, but rather as instinctual emotional reactions that are deeply ingrained in our psyches as a result of natural selection.

Love and imprinting

Because marriage is not synonymous with love, this does not mean that love does not exist. In fact, most of us do experience a highly intense emotional feeling directed towards another person at some time in our life – perhaps several times – that we feel is only adequately described by the word 'love'. Although it seldom lasts without attenuation, or at least modification, for a lifetime, it is nevertheless of great importance in our lives and to human relationships in general.

Some sociologists have tried to dismiss love as the excuse we use for changing our life style – leaving our parents to live with another person, getting married, or getting divorced. It is true that love frequently is accompanied by some transition of the kind, but this does not explain away the phenomenon entirely. It appears that, like chimpanzees, we do have a biological tendency to pair off with a particular individual (usually of the opposite sex) for a period of concentrated and exclusive sexual activity that frequently leads to conception (and would do a great deal more often in the absence of birth control techniques). Usually this love attachment begins with a degree of active courtship on the part of the male, but it is not long before it is fully reciprocated by the female. Termination usually occurs when the female finds the male insufficiently attentive and switches her romantic interest and devotion to another more ardently wooing male. Men very seldom break off affairs decisively; they let them drift until the woman can take no more and vows to reassign her affection. As we have noted, such reassignment is much more likely to occur if the relationship has not borne children.

The most extreme form of pair-bonding in the animal world is seen in birds, and it has many characteristics in common with what we call *imprinting*. This is the highly inflexible attachment of a new-born baby bird to the first moving object that remotely resembles its mother. In the natural world this is quite adaptive because the first moving object to greet the eyes of a newly hatched bird nearly always is the mother, but Konrad Lorenz has shown that imprinting can occur in relation to moving models of the mother and even human beings wearing boots of a certain size and colour. During one period of his research, Dr Lorenz found it difficult to go anywhere without being trailed by baby geese. Once imprinting has been established it cannot easily be altered; the young bird will follow the object of its attachment even to its own destruction.

The psychologist John Bowlby (1971) has reviewed a great deal of research which suggests that human children imprint upon their parents during early infancy and that this is an important basis of human attachments in general. Although the tendency for the child to cling to its mother is most obvious between the ages of six weeks and two years, remnants of this need for bodily contact are seen throughout adolescence and adulthood and even into old age. At all ages the need to cling to another person increases when the environment becomes threatening in any way, so Bowlby supposes that the main function of attachment is protection from predators. When a monkey troupe is threatened by enemies, the babies run to their mothers and cling to their fur so that they can be transported out of the danger area. Even in adulthood it is advantageous to stick with other members of one's species, particularly when danger threatens. There is always safety in numbers.

At one time, when American environmental behaviourism dominated the psychological scene, it was widely believed that attachment was a 'secondary drive' which develops after some experience with the mother-figure as a supplier of food. (The term mother-figure is used in preference to mother, because it has always been recognized that attachment can occur to any adult, whether or not she is the biological mother.) However, in recent years, the observations of Lorenz on birds and Harlow's (1965) experiments with infant monkeys have established fairly convincingly that young animals have an instinctual need for contact with a stable mother-figure even if the food supply comes from a totally independent source. Thus it seems that the tendency toward attachment with one special other animal (normally of the same species) is instinctual; all that is learned is the precise target of that attachment.

The extent to which the phenomenon of romantic love in adulthood derives from this parent-child attachment remains uncertain. Probably adult love is 'multi-factorial' – an outgrowth of several different basic instincts. One of these predisposing instincts is no doubt the need of the infant to seek protection from its parents, or some substitute. Thus people will take comfort from friends, lovers, the spouse, or even an abstract idea such as God, particularly in times of stress and anxiety. But there is no doubt some contribution from the other side of the coin – the parental protection instinct. Men like girls who emit infant signals like big eyes and soft skin; they describe their girl friends as cute and cuddly, use diminutive nicknames, and indulge in baby-talk when being affectionate. Women very often enjoy the 'little boy' aspects of their boy-friends and husbands, however much they may superficially deride them. 'I have to hang up all his clothes for him,' they tell their friends with barely disguised pride. They too may address their man as 'baby' and bestow motherly tenderness and care upon him. No observer can fail to be impressed by the way in which lovers appear to regress to childhood in various aspects of their behaviour – holding hands, kissing, cooing and grooming. In a sense they take it in turns to be parent and child.

Bowlby points out that infants develop not only a protective mother-figure to whom they run and cling when distressed, they also attach themselves to special playmates. Playmate attachments are distinguishable from the mother attachment in that they occur slightly later in life and the playmate is sought in times of health and good spirits rather than during illness or stress. This, then, is a second kind of imprinting that is probably basic to some aspects of adult heterosexual love. Particularly in the early stages of courting, many men and women prefer to present their better and brighter self to their lovers, and withdraw to solitude or their family when feeling off-colour. This selective self-presentation functions, in the short term, to create an image of a suitor who is attractive, healthy and good company. But insofar as it is a form of deception it may eventually result in a feeling of disillusionment on the part of the partner which contributes to the general process of disenchantment that spells the end of romantic love.

A third kind of imprinting that has been identified in animal studies, different again from either of the previous two, and almost certainly under separate hormonal control, is sexual imprinting. There is still some question as to when this occurs in the life-span, but there seem to be important periods around the ages of three and four in humans and again in adolescence. In early childhood, it seems, major choices of sex objects are made. It is probably around this time that sex orientation (homosexual versus heterosexual) is largely fixed, as well as a number of other preferences such as fetishistic attachments (Gosselin and Wilson, 1980). In adolescence the sex hormones become highly active again, this time to increase libido in general, and to determine specific romantic and sexual interests of the kind that (from our probably jaundiced adult perspective) are likely to be called 'puppy love' or infatuation.

A detailed questionnaire study of adult love styles by Canadian sociologist John Lee (1973) points to the primacy of three main types of love. These he called *storge*, *ludus* and *eros*, but more descriptive terms would be companionate, playful and passionate love styles. Any given individual may have a general preference for one type of love more than the others, but most people develop relationships corresponding to each type simultaneously or sequentially, or seek to satisfy all three needs in the one partner. Probably the difficulty in achieving the latter is one of the reasons why so many marriages collapse in Western society. Most of us have needs for all three kinds of love and it is almost impossible to satisfy them by an exclusive attachment to one other person. Those people who do report total satisfaction in a monogamous relationship may have succeeded in doing so, and if this is reciprocated and lasting then so too will be the marriage. Other people may have love needs so focussed on one corner of the Lee triangle that conflict does not arise, and this would also explain satisfaction in a monogamous relationship. For example, some people may be so low in libido, and so 'mature', that the erotic and ludic components become almost irrelevant and a single relationship that satisfies the need for companionship may be sufficient to ensure happiness. This is probably the case with a lot of contented elderly couples.

The interesting thing about the broad three-way classification of love styles given by Lee is how nicely it corresponds to the three main types of childhood attachment identified by Bowlby. The first infantile attachment, that to the mother, which serves as a kind of security buffer, seems to have echoes in the companionate love style (*storge*). This kind of love is often most pronounced in middle and later years, perhaps because of increasing loss of social mobility with age, and fears relating to the approach of death. Perhaps also, the other love needs are proportionately diminished along with libido because sex hormone levels in the blood are sharply reduced in middle age.

The second attachment of the child, that to his playmates,

with its exciting, exploratory nature, which is most enjoyed in times of health and safety, seems to correspond fairly well with Lee's playful type of love (*ludus*). Obviously, this kind of love is more prevalent and apparent in younger people, although there are people who (delightfully) 'never grow up'. Playmates are usually more readily replaced than mother-figures; likewise the ludic lover moves easily from one partner to another with minimal heartbreak (indeed, variety may be enjoyed for its own sake).

Finally, Lee's erotic love style seems to be a direct derivative of sexual imprinting, with its strong sexual component and emphasis on the visual aspects of the love object. In this type of love, more common in men than women (Hatkoff and Lasswell, 1979), there is an almost fetishistic attachment to the face and form of the loved one. Replacements are not acceptable, and considerable distress is occasioned by a break-up.

A word about the question of maturity: In everyday language we refer to certain adult attachments as 'immature', the word being used in a derogatory way rather as 'infatuation' is. We point to a small man who has married a large woman and say that he has acquired her as a mother-figure; she in turn is accused of 'mothering' him. The forty-year-old playboy bachelor who enjoys sex with a variety of beautiful girls in their twenties is described as immature because he has not 'settled down' with a wife and children. Somehow we manage to convince ourselves that there is something inadequate or pathological in these behaviour patterns and that the individuals concerned must be discontented at some deep level or other. Possibly they are, but then how many of the rest of us can boast total contentment with complete honesty?

There is no real reason why one instinct basic to adult heterosexual attachments should be regarded as more admirable than another. Folk wisdom does seem to have correctly identified some of the developmental origins of love styles – especially those that connect with the mother-child relationship and play – such that there is a sense in which the concept of maturity is relevant to sexual behaviour. However, the love style that is usually lauded as the most 'mature' – the companionate form of love – could well be the most infantile in origin of them all.

While the evidence is not conclusive at the moment, there are indications that the absence of a stable mother-figure in childhood can result in difficulties in establishing normal social relationships in adulthood. Harlow's monkeys, when reared in the absence of a mother or any satisfactory, cuddly alternative, remained neurotic and isolated throughout their lives (Harlow & Harlow, 1965). Likewise, the failure of infant monkeys to form normal playmate relationships was usually followed by failure to develop adequate sexual behaviour in adulthood. In this respect the sexual performance of the male monkey was rather more sensitive than that of the female, perhaps because the male sex role requires more social confidence and skill than that of the (relatively passive) female. This is consistent with the fact that sexual deviation is much more common in human males than females.

Despite a great deal of folk lore relating to the effects of broken homes, there is no evidence that the absence of a father during childhood results in sexual problems for either boys or girls. While the father does seem to be involved in the development of gender identity to some extent making boys more masculine and girls more feminine (Mead & Rekers, 1979) his absence does not lead to major sexual difficulties. In a recent study that I conducted with the help of a national newspaper, it was found that children who grew up with little or no contact with their fathers actually reported less interest in homosexuality and fewer sex problems than children coming from normal two-parent homes. It seems that the capacity to love does not require a stable father-figure in addition to the mother. One loving parent is probably sufficient for normal development. But note: while we usually call this one parent a 'mother-figure', for all we know at the moment a man might do just as well (Morgan, 1975). These questions will be taken up in a later chapter in connection with the topics of deviant and dysfunctional sexuality.

Fixing on the love object

If imprinting processes dating from childhood are of major importance in the origins of love, this should have a profound effect upon the characteristics of the person that we fall in love with. We would expect that various aspects of people (their eye colour, hair style, figure, voice and so on) would appeal to us or put us off according to the nature of our early experiences with people who share these characteristics. Imprinting theory would suggest that we build during childhood a blueprint of the kind of person that we are prone to fall in love with in adulthood. Individual differences in early experience would then account for seemingly peculiar preferences that are referred to as the 'chemistry of love' – emotional imperatives that pay no homage to the objective indications of suitability of the kind applied by computer matching bureaux.

An interesting implication of this theory is that when people replace their mate following separation or bereavement, they will tend to choose another partner with similar characteristics – even if this means repeating their apparent mistakes. A highly intelligent university teacher that I knew in the United States was divorced from a rather large lady whose excessive weight I had presumed to be one of the reasons for his unhappiness with her. I was therefore intrigued upon meeting his second wife to discover that she was equally immense. Apparently, fatness was an attribute that he found attractive in women; it was not the turn-off that it is generally supposed to be. British sports writer David Emery created a great deal of publicity when a few years after his fiancée, the athlete Lillian Board, had died of cancer, he married her twin sister. This could be thought of as an ideal way of coping with the loss of an imprinted love object. Identical twins are as close as we come in real life to replicas of people. I do not wish to demean the behaviour of any of the parties involved in these relationships; I merely use them as illustrations to support the idea that in falling in love with someone we may be imprinting on certain visual or other sensory aspects of them, or carrying over an earlier imprinting of this kind.

Sensory imprinting, particularly the visual kind, seems to be more typical of men than women. Although no direct proof is available at the moment, a reasonably plausible theory that might account for this concerns the importance of right hemisphere specialization in men. On average, men seem to have a right side of the brain that is highly developed for the processing of visual patterns – spatial relationships as they exist at one moment in time. In both men and women, the left hemisphere deals more with verbal and logical processes that are arranged sequentially in time. Now there is a sense in which, compared with men, women appear to function as though they have two left hemispheres. The result is that men are usually superior at the mental representation and manipulation of spatial arrangements (such as map-reading and chess-playing) while women are often superior at tasks requiring verbal memory and time orientation (like remembering dates and telephone numbers).

One of the ways in which this gender difference in cognitive functioning seems to show up is in conceptions of love. Men are more likely to fall in love with the visual features and configuration of the love object, whether this be a woman, another man in the case of homosexual love, or even an inanimate object as in the case of fetishism. For a woman, the definition of love is much more bound up with memories of the nice things the partner has said and done, and expectations concerning the future of the relationship. Of course, this difference supports the optimal reproductive strategies of males and females, the male being sexually attracted to present targets and the female being more concerned with satisfaction and security in the long term.

Women are also prone to fall in love with the man who initiates them sexually. In mythology, an undercurrent of this predisposition is recognized in the 'awakening' themes of Sleeping Beauty and Wagner's version of the Siegfried legend. Dracula marks his territory and claims female souls with the sexually symbolic act of neck-biting and penetration. A related phenomenon in real life is the tendency for women to fall in love with men who have abducted, imprisoned and raped them, the most famous recent example being Patricia Hearst in California. Falling in love with the man who has conquered her, whether honourably or by brute force, serves the woman's general instinctive strategy by motivating her to try to persuade the father of her possible offspring to stay and support her.

Family resemblances

Freudian psychologists have long maintained that in choosing a marriage partner we are in some sense looking for a replacement for our opposite-sex parent. The classical Freudian theory is that around the age of five little boys fall in love with their mother and little girls become passionately attached to their father (the so-called Oedipus and Electra complexes). As they grow older, children learn that sexual attraction to their parent is unacceptable (not least because their same-sex parent is likely to become dangerously jealous – the so-called 'castration complex') so they 'resolve' their attraction in some way, usually by 'repressing' their Oedipus or Electra complex, relegating it to the 'unconscious mind'. There it stays, for the rest of their lives, though it may still manifest itself in various indirect ways such as influencing their choice of marriage partner.

This theory of the Oedipus complex is not highly regarded by modern psychologists (except for the minority who follow the teachings of Freud word-for-word, with near-religious devotion) but it does seem probable that our attitude towards our opposite-sex parent has some part to play in determining our response to the opposite sex in general, and our preferences among them. One experimental study of this question concerned preferences among different male physiques in a sample of girls. Alan Miller of California State University, Los Angeles (1969) had the girls pick out from a set of silhouettes the male figure that they found most desirable as a mate and also the male figure that was most similar to their father. He found a tendency for the girls to prefer as a mate either the figure that was most like their father or the one that was most *unlike* their father. While hardly confirming the Freudian theory of the Oedipus complex, this finding does strengthen the idea that a girl's attitude towards her father is one determinant of her choice of husband.

Perhaps the best evidence so far available in support of the Freudian idea of mate selection comes from a recent study by Jedlicka (1980). He studied a sample of nearly one thousand brides and grooms of mixed ethnic parentage who had been married twice in Hawaii – an area with a high proportion of racially mixed marriages. What Jedlicka found was a fairly striking tendency for people to choose a spouse from the same racial group as their opposite sex parent. Men married into the ethnic group of their mothers and women married into the ethnic group of their fathers. This pattern of preferences applied with equal strength for men and women and for first and second marriages. Although other explanations are possible, this finding is consistent with the idea that we carry over into adulthood some remnants of our early love for our parent of the opposite sex.

The British psychiatrist Dr John Birtchnell (1979) tested another psychoanalytic idea – that our relationships with our siblings (brothers and sisters) have some influence on our compatibility with different partners. In particular, he was interested in the idea that strong heterosexual relationships sometimes develop between brothers and sisters and that these form a model for the kind of love relationship that we seek in adulthood. For example, we might expect that a man with an older sister would be inclined to pair off with a woman who grew up with a younger brother, and vice versa. In addition, we might find that such partnerships are happier and more stable, since each individual would bring to the relationship appropriate experience in relating to the opposite sex. Birtchnell interviewed two thousand couples, half whose marriages were successful and half whose marriages were judged by themselves as unsuccessful. Unfortunately for the theory, he was not able to discover any mating patterns that repeated the subjects' own positions among their siblings, nor was there any difference between successful and unsuccessful marriages in this respect. Only one positive finding emerged: among a subgroup of twenty-seven per cent of his total sample who expressed a special fondness for a particular sibling, there was a tendency for men who were particularly fond of an older sister to have married women who only had younger brothers. Birtchnell concluded that, overall, sibling relationships do not have a major effect upon marital choice, but for a few particular individuals they may be highly significant.

Assuming there is some element of truth in the general idea that family likenesses are in some way involved in our choice of love object, the imprinting mechanism would seem to offer a reasonable explanatory model. Human babies develop strong attachments to particular cuddly toys which persist well into later childhood however dirty and ragged the object becomes. Fetishisms relating to inanimate objects and materials such as high-heeled shoes, rubber or fur also seem to start in childhood but persist, and may even be magnified, in later years (Gosselin & Wilson, 1980). If we are capable of forming such inflexible and seemingly irrational attachments, then it seems likely that we will also grow up with predispositions to fall in love with people displaying particular features such as widely spaced eyes, a cleft chin, or subtler aspects such as facial expressions and tone of voice. Some of these will be positive, attractive criteria; others will be negative, or 'aversive', leading to idiosyncratic avoidance of potential partners.

Many people report having had the experience of feeling an emotional reaction towards a new acquaintance that they cannot explain to themselves for some time, until suddenly one day (or night) the realization strikes them that this person reminds them of a member of their immediate family or some other person of great significance in their childhood. At the same time their introspection tells them that the emotion (positive or negative) that they feel towards the new person is in some way derived from that appropriate to the other party. Thus they feel empathy with the girl who reminds them of their sister, or fear of the person who is reminiscent of an old tyrannical headmaster. The process is akin to that which produces a conditioned anxiety reaction to the smell of a dentist's surgery, even if we are only there to have a check-up or pay the bill. This then, is the kernel of truth in the idea of the Oedipus complex; experiences with our parents lead to certain indelible attractions and repulsions which contribute in subtle ways to our adult preferences.

Of course, when the psychoanalysts say that we are trying to replace our parents in our marriage partners, they are probably referring to social roles as much as physical looks. The man is no doubt looking for a woman who will provide the nurturant care that was bestowed upon him by his mother as a little boy, so he expects that she will cook his meals, wash his clothes, be sympathetic, and so on. The woman is probably seeking in her husband a fatherly kind of strength and protection – a man powerful enough to sweep her off her feet figuratively if not literally. At the very least he must be able to open jam jars and mow the lawn. Many marriages fail when we discover that our partner is unable or unwilling to fulfil these assigned roles. Thus a recent survey which I conducted through *She* magazine showed that women who reject the traditional female role and who are emotionally involved in the feminist movement are less likely to be happy in their marriage. While cause and effect are a little difficult to untangle here, it is clear from the preceding discussion that the idea of a male-female partnership is based not just on sexual compatibility but also on a symbiosis of social roles.

The shift towards a unisex society may result in greater justice, but it may also turn out to be the final blow for the institution of marriage, which has already been badly shaken by a succession of social changes such as urban anonymity, geographical mobility, and efficient birth control. While 'falling in love' does seem to be a biological imperative, marriage is not. In many ways it is antithetical to human nature and its future is therefore uncertain.

6 Sexual Responsiveness

We have seen that there are considerable differences between individuals and between the sexes as regards the speed and ease with which they become aroused. In general, men are more readily aroused than women, especially by visual stimulation, and women may be rendered more arousible by the injection of male hormones. This tendency for men to be 'turned on' quickly by the presence of attractive females was seen to support the male 'promiscuity' strategy. By the same token, it would be inappropriate for women to be easily aroused by the sight of a man, since they would lose their power to control the occasion for reproduction.

Males and females also differ in the ease with which they can achieve satisfaction during sexual intercourse. Once intercourse has commenced men are usually guaranteed an eventual feeling of completion through orgasm. Women, however, are frequently left 'hanging' and wishing for more, which the exhausted man may be unable to provide. This apparent injustice has been a major complaint of the feminist movement who tend to blame men for it. Chauvinism, they maintain, has been carried into the bedroom. If men were sufficiently sensitive and considerate as lovers there would be no such problem. But is the problem as simple as this? Perhaps there are also some biological differences between men and women that would make orgasm more difficult for women. Several psychologists and ethologists have addressed themselves to this question and come up with some interesting theories.

Nature of the female orgasm

The female orgasm may be described as a highly variable peak of sexual excitement accompanied by rhythmic contractions of the outer third of the vagina (and often also of the uterus, rectal sphincter, and urethral sphincter). This is immediately followed by release of the vasocongestion and muscular tension which has accumulated over the course of the sexual exercise. According to Masters and Johnson (1966) the physiological responses of men and women during orgasm are about as parallel as they can be, given the anatomical differences. For example, the contractions in the outer part of the vagina are timed at the same intervals as the contractions of the penis which propel seminal fluid through the urethra during the man's ejaculation. There are two interpretations of this fact: one is that men and women are constructed for sexual compatibility – so that their sexual responses will coincide as far as possible, leading to harmony in love-making and the possibility of stable marriages. The other is that the similarities are a simple function of the fact that men and women share the same anatomy and physiological function unless there is some compelling evolutionary reason why they should not. We will see that the latter theory fits better with the overall facts.

The main difference between men and women as regards orgasms is in the reliability with which they occur. Men very seldom have difficulty in achieving orgasm, but for women the climax is notoriously unreliable. Some fairly typical figures are those provided by Tavris and Sadd (1977) in their *Redbook* survey of a hundred thousand American women. About fifteen per cent of women claimed that they 'always' had orgasm during intercourse, 48 per cent said they achieved orgasm 'most of the time', 19 per cent said 'sometimes' and 11 per cent 'once in a while', while 7 per cent 'never' experienced orgasm during intercourse.

More detailed studies of female orgasm suggest the importance of clitoral stimulation. For example, Fisher (1973) reported that among his 38 per cent of women who 'always or nearly always' had orgasm during intercourse, 63 per cent used clitoral stimulation before intercourse and 35 per cent stimulated their clitoris (or had their partner do so) during or after intercourse. Quite a high proportion of women who claimed to be fairly reliably orgasmic admitted that direct manual stimulation of their clitoris was usually necessary; intercourse by itself was not generally enough. Hite (1976) found in her survey of 3000 women that 30 per cent were able to have orgasm during intercourse without manual stimulation, while an additional 19 per cent could reach orgasm with manual assistance. The other 51 per cent rarely or never achieved orgasm during intercourse, although quite a lot were successful with masturbation. Of the 30 per cent who could have orgasm during intercourse without manual help, most were found to favour coital techniques in which their clitoris was stimulated (for example by their partner's pelvic area) in addition to penile thrusting.

Western sexual mythology maintains that women need a great deal of foreplay, and in our culture at least, this usually involves a certain amount of clitoral stimulation. Survey figures suggest that foreplay typically lasts for about 10 to 15 minutes, while intercourse itself usually lasts about another 2 to 10 minutes (Symons, 1979). In a recent survey of 4000 British people that I conducted with the help of a national newspaper, 'insufficient foreplay' was the most frequent criticism women had to make about their sex lives; nearly one-third raised this as their primary complaint. Apparently the human female requires a fairly sustained period of stimulation, particularly in the clitoral region, if she is to experience orgasm.

Cultural variability

Margaret Mead (1961) reports a fairly astonishing amount of variation from one culture to another as regards the incidence of female orgasm. In a few cultures all women are said to have orgasm. In many others, none of the women are supposed to have orgasms and even the concept of female orgasm may be unknown. Most, though, are (like our own culture) somewhere in between; the possibility of female orgasm is recognised but its occurrence is known to be somewhat unreliable. Most of the women have orgasm some of the time, and some of the women have orgasm most of the time.

Whether or not female orgasm is common in a given culture seems to depend upon the extent of general sexual repression prevailing in that culture and the extent of male skill in love making. An absence of the former and a high degree of the latter seem to be necessary, though perhaps not sufficient, conditions for the widespread appearance of female orgasms (Symons, 1979).

In Mangaia, one of the Cook Islands in the Pacific, boys aged thirteen or fourteen are given instruction in sexual techniques, including breast-kissing and cunnilingus, and are taught the importance of delaying ejaculation until the girl has had several orgasms (Marshall, 1971). Two weeks later there is a practical exercise in intercourse with an older, experienced woman, who teaches the teenage boy how to delay ejaculation and time it so that it occurs simultaneously with his partner's orgasm. At the same time, the girls are also given verbal instruction by older women. The goal of this sex education is to achieve intercourse lasting fifteen to thirty minutes with continuous thrusting and active female participation during which the female has two or three orgasms. The final orgasm is supposed to be simultaneous with that of the male partner. Mangaians believe that female orgasm must be learned with the help of a good man. If a man fails to bring his partner to orgasm she is likely to leave him for someone else more satisfying, thus ruining his reputation among other women. In this culture, there is not only a high level of permissiveness but a positive concern for female satisfaction and the development of expertise. The result is that female orgasm is widely enjoyed.

In the Melanesian island district of 'East Bay' studied by Davenport (1965), female orgasm was achieved quite regularly by extended mutual masturbation, with insertion of the penis just before both partners reached orgasm. Again, it seems to be freedom from inhibition combined with highly developed male skill that enables the female to enjoy orgasm.

However, these Pacific peoples are quite exceptional within a world context. In the vast majority of societies, men take the sexual initiative and proceed quickly to their own orgasms with little regard for the response or enjoyment of their female partner. The result is that women in these societies very seldom experience orgasm. Some cultures, particularly in north and east Africa, take active steps to prevent the likelihood of female orgasm by the operation of clitoridectomy – surgical removal of the clitoris. But even in cultures that do not perform this operation, female orgasm is a fairly rare phenomenon.

The animal evidence

Although there has been some disagreement over the extent of female sexual response among non-human animals, it becomes progressively clear that they do not often experience orgasm in the normal course of events. Female mammals are by no means totally passive and submissive; they do demonstrate proceptivity (that is, they seek out and presumably enjoy sex, especially when they are on heat) but there is no compelling evidence that they have orgasm in the human sense. They sometimes display excitement, even crying out, but a clear climax followed by sudden relaxation is not observed in the female in quite the same way as it is in the male.

Evidence for orgasm in primates is equally unsatisfactory, at least as it applies to animals in the wild. The only real evidence for primate orgasm comes from zoos and laboratories, where it appears that a degree of stimulation was applied that would not occur naturally. For example, Burton (1970) harnessed three female rhesus monkeys on a frame, fed them, groomed them until they presented, and then gave them five minutes of clitoral stimulation. This was followed by five minutes of vaginal stimulation with an artificial penis, four minutes of rest, and then another five minutes of vaginal stimulation. In these circumstances, two out of three of the monkeys showed vaginal spasms indicative of orgasm, but the degree of attention given by Burton cannot be compared with the usual three or four seconds for which rhesus copulation typically lasts.

In captive monkeys female orgasm is most likely to occur as a result of 'lesbian' contact. Sometimes when a female mounts another female she will make pelvic thrusts that culminate in muscular spasms suggestive of orgasm and produce facial expressions and vocalizations similar to those of the ejaculating male. The same does not happen to the female who is mounted, whether by another female or a male. This observation calls to mind the finding of researchers like Hite that human females are often better able to obtain satisfaction from another woman than from a man. The reason presumably is that they continue clitoral stimulation for a longer period of time without the termination that tends to occur following a man's orgasm.

Overall, the evidence reviewed by Symons (1979) suggests that orgasm in female animals can only be obtained with prolonged clitoral stimulation of a kind that would very seldom occur in nature. Even then, it appears highly unreliable; not all females experience it, and those that do are not consistent in their response.

Is orgasm essential to female enjoyment?

The current emphasis on women's orgasms seems to imply that sex without orgasm is either useless to a woman or actually unpleasant. Apparently this is not necessarily true. Some women do find sex without orgasm worthless or frustrating, but an almost equal number of women report that it is quite enjoyable regardless of whether or not orgasm occurs (Wallin, 1960; Shope, 1968; Sigusch and Schmidt, 1971). Some women may even find orgasm unpleasant; a frightening or humiliating experience. Hite's (mainly feminist) sample of women cited feelings of intimacy with their partner as the prime reason for enjoying sexual intercourse, and when they were asked about their favourite physical sensation during sex, the moment of penetration was given top billing more often than orgasm.

It seems, then, that orgasm is not the 'be all and end all' of female sexuality. This is fairly consistent with the animal evidence. Most primate females seem to enjoy sex when they are on heat, and they tend to be indifferent to it when they are not. They seldom show distress when intercourse is brought to an end when the male reaches orgasm, and although they do not appear to have orgasm themselves, they show no evidence of frustration nor any burning desire for further intercourse. They may even move away from the male as if to avoid a repeat performance. For example, Nadler (1977) describes the postcoital behaviour of orangutans as follows: 'Following separation, the male turned away from the female, reclined in a prone position, and peered out of the cage, groomed his genitals or went to sleep. The female generally climbed to the ceiling of the cage and examined and groomed her genitals.' Unless the female orangutan is seeking some simian equivalent of 'sex hanging upside down from the chandelier,' it hardly sounds as though she is anxious to recommence action.

It is very doubtful that orgasm is the natural, preordained comsummation of sexual intercourse for the human female. As Mead (1967) observes, 'that whole societies can ignore climax as an aspect of female sexuality must be related to a very much lesser need for such climax.' While it is undoubtedly true that frustration can be experienced by women if intercourse is terminated abruptly at a point when they are very close to climax, this is probably a result of the long build-up and the high degree of excitement to which they have been brought. If women have a 'quickie' of the kind that female apes are usually treated to by their chauvinistic male suitors in the wild, no great degree of frustration is usually felt. Similarly, women who have never had an orgasm do not necessarily miss it, and if they do, it is usually because they have read so much about it in books and magazines. The greater part of female sexual frustration, of the kind stemming simply from failure to reach orgasm, is probably of recent origin.

Theories of the female orgasm

This brief review of the evidence relating to the occurrence and importance of female orgasm in various cultures and species provides a background against which to consider theories concerning the evolution of women's orgasm.

The first theory to be considered may be called *repression theory*. This is stated most explicitly by Sherfey (1972), although similar ideas have been expressed by the European philosophers Marcuse and Reich. According to this view, female orgasm occurs easily and reliably in non-human animals and its unreliability in the human female is a result of social pressures and restrictive learning. Women are different from female animals in that they are taught by the repressive forces of civilization to control their sexual urges. Women are presumed to be naturally sexy and promiscuous, possibly even more so than men, but control has been politically imposed upon them in the interests of family life and social stability.

Repression theory strikes a chord with the feminist movement by blaming sex roles for women's difficulty, and perhaps for this reason, has attained a great deal of popularity in recent years. It is, however, almost certainly wrong, since it conflicts with virtually all the known facts. Female orgasms are not more common in the animal world than the human world; it is doubtful if non-human females have orgasms at all, and if they do, they are certainly very infrequent. The females of primitive tribes, who are presumably relatively free of the repressive effects of civilization, are no better off than 'modern' women as regards frequency of orgasm. Apart from a few selected Pacific cultures, such as those described above, the modern European woman is more orgasmic than most. Learning does seem to enhance the possibility of female orgasm, but it seems to be a matter of positively learning a skill rather than reducing culturally imposed inhibitions. Finally, the evidence reviewed in preceding chapters is counter to the idea that women are naturally libidinous to the same extent as men or more. Rather, it appears that a certain amount of caution and control is natural to the females of most primate species and perhaps especially the human female. In order to become orgasmic and fully enjoy sex some women may need to learn to override this control, but the inhibition is mainly of biological origin not political or social expediency.

A second theory of the evolution of female orgasm is that of *pair-bond consolidation*. This theory has received a great deal of publicity through the writings of Desmond Morris, but much the same idea has been propounded by workers in a variety of different disciplines including psychologists (e.g. Beach), psychiatrists (e.g. Hamburg), ethologists (e.g. Eibl-Eibesfeldt) and biologists (e.g. Barash). These theorists believe that female orgasm is much more highly developed in humans, if not necessarily unique to them, and that it has evolved to promote monogamy and make family life more rewarding. As mentioned in the previous chapter, humans are seen as having arrived at a bird-like pair-bonding arrangement through a separate evolutionary channel. Female orgasm and the loss of oestrus are both construed as mechanisms which support this development. If the human female is sexy at all times of the month and has very rewarding orgasms along with her male partner, this will help ensure that copulating couples fall deeply in love and thus stay together to cooperate in the task of child-rearing.

This theory is at least consistent with the fact that human females seem to be more sexually responsive than non-human females, but there are other problems with it. If female animals never had orgasm in the natural state it is very difficult to see how orgasmic ability could have been selected for in the first place. It would need to have emerged quite suddenly and late in the evolutionary time scale having been accidentally introduced (presumably by mutation) as a reasonably common variant of female response. This being the case, it is somewhat strange that our nearest cousin among the great apes, the chimpanzee, is about the least female-orgasmic of them all. We would also need to ask why, if the female orgasm is advantageous to women (in the evolutionary sense), it has not evolved a more surefire operation. In fact, there is no real evidence that orgasmic women have any breeding advantage over non-orgasmic women.

Another problem with the pair-bond consolidation theory is the unsatisfactory evidence to support the rather idealistic view that the human ape is more monogamous than its cousins. Evidence discussed in earlier chapters suggests that we are basically inclined towards polygyny, just like other ape species in which the male is bigger in size than the female. While we do fall in love and become exclusively absorbed with one other person for certain periods of time, this seldom lasts long after the arrival of children, and despite socially imposed marriage obligations, husbands as well as some wives, remain sexually exploratory to a greater or lesser extent.

We have noted previously that the strength and duration of the human bond displays no simple relationship with the frequency of consummation. While sex with a particular man may increase a woman's desire to stay with him (in accordance with her sociobiological need for his protection in the event of pregnancy) repeated sex with the same woman usually increases the strength of a man's wanderlust (in accordance with his potential genetic advantages). Conversely, human love is often known to grow when the lovers have to part for some time. Altogether, it seems that the pair-bond consolidation theory of female orgasm espoused by Morris and others is not well supported by the facts.

A third theory of female orgasm that has been around in various forms for a long time is described by Wilson (1979) and may be called the *jackpot theory*. According to this idea, there are two main functions of orgasm; (1) to reward people for engaging in sex, and (2) to terminate intercourse at a suitable juncture, rather than have the couple pump on forever. While the first function of orgasm (reinforcement) might seem beneficial to the reproductive chances of a woman because it would encourage her to have sex more often, the second function (termination) would be detrimental if it meant that intercourse was often interrupted before ejaculation. Evolutionary forces have therefore arrived at a female orgasm which compromises between these conflicting pressures. It differs from the male orgasm in that it does not occur so quickly or decisively, so that intercourse is not likely to be terminated by the woman before she has been inseminated. Nevertheless, high points of ecstacy may still be obtained, sometimes several times in succession, so that sexual activity is well rewarded. In Skinnerian terminology, women are on an 'intermittent reinforcement schedule', which is notoriously slow to extinguish. An analogy is seen in the person who spends long hours pushing coins into slot machines even though small prizes are infrequent and jackpots extremely rare. By contrast, men are on a hundred per cent reinforcement schedule, which is known to lead to a high rate of repetition of the behaviour in question. It would extinguish relatively quickly if reinforcements were suddenly withheld, but of course they very seldom are – intercourse is nearly always sustained until male orgasm.

While this theory accounts satisfactorily for the equilibrium that seems to have been arrived at by the human female, it does not adequately explain the late emergence of female orgasms in the evolutionary tree. Like the pair-bond consolidation theory, it stumbles on the fact that female orgasms do not seem to have occurred often enough in the wild for their selective advantage to have taken a hold. The problem is not so much in explaining why they are slow and less decisive but in understanding why they occur at all. It might be supposed that a sudden change in anatomy occurring between our primate ancestors and ourselves paved the way to female orgasms, but this is rather belied by the fact that other apes do seem to have the physiological apparatus and capacity for female orgasm without capitalizing on it. Since the mechanism for orgasm is present in female chimpanzees it is hard to see why the rewarding function was not discovered and established. Enjoyment of orgasms seems fairly unnecessary to the reproductive success of female chimpanzees, so why should women be any different?

This leads us to the fourth theory of the female orgasm, called *artifact theory* (Symons, 1979). Symons proposes that the female orgasm has no evolutionary advantage at all, no matter how much pleasure it may give to women who experience it. He believes that women (like female mammals and primates) have a variable capacity for orgasm which is a by-product of the fact that they share a great deal of common neurology with the males of their species. This follows from the general principle of evolutionary sex differentiation, that males and females will be constructed identically unless there is some compelling reason why they should not (presumably because this economizes on genetic coding problems). Thus the basic plan of a human being is bisexual in its potential. If the individual develops as a female, the breasts are enlarged and elaborated so as to become fully operational but the clitoris remains small and hidden. If the foetus develops as a male, the clitoral area is expanded into a highly functional penis but the breasts remain perfunctory. The erotic sensitivity of female breasts has presumably developed in the service of the suckling instinct, and therefore there is no real necessity for men to obtain erotic pleasure from their nipples (even though some do). Similarly, orgasm is the highly pleasurable experience accompanying the muscular spasms which project seminal fluid into the vagina and which compel the male to stay *in situ* until completion. There is no equivalent biological significance of orgasm for the female as far as we are aware at present.

Thus Symons argues that the female orgasm is not 'natural' in the sense of having any evolutionary function or purpose. Rather it is a capacity which, like playing the piano, can be learned by a high proportion of women and which may provide the performer with a great deal of pleasure. This does not devalue either activity; if anything it emphasizes their common status as higher cultural achievements rather than biological necessities.

Are female genitalia designed for orgasm?

One major implication of artifact theory is that female genitalia are not designed for orgasms any more than fingers have evolved to play the piano. The research of Masters and Johnson has established that nearly all female orgasms (however caused) are triggered by clitoral stimulation. But during intercourse this is nearly always indirect – a result of pushing and pulling the labia. Nevertheless, Masters and Johnson seem to imply that this kind of stimulation ought to lead naturally to female orgasm. Such a conclusion is almost certainly erroneous and comes about as a result of two biasses that may be identified in their work: (1) their requirement that all female subjects were orgasmic, which would lead to a sample weighted with fairly libidinous women, and (2) their explicit enthusiasm for marriage, which seems to entail a presumption that male and female sexuality are perfectly complementary by nature. These factors apparently led Masters and Johnson to argue that male and female genitals are equally adapted to orgasm during intercourse. This, of course, does not correspond with the evidence that many women find masturbation, either by themselves or their partners, the superior route to orgasm. Both Kinsey and Hite found that women can more easily reach orgasm through masturbation than through intercourse, and Masters and Johnson themselves found that the *best* female orgasms (whether measured by subjective report or the strength and number of contractions) occurred during masturbation not intercourse.

Another interesting comparison is in the devices and techniques of masturbation used by men and women. When men masturbate they usually simulate intercourse as far as possible, occassionally to the extent of using inflatable dolls with operational vaginas. Phallus-shaped vibrators are available for women, but they more often place them around the area of the clitoris rather than insert them into the vagina. In fact, the vibrators that most women find best are not penis-shaped at all but are flat-headed and driven by mains supply electricity. Clearly, deep penetration of the kind that is obtained from intercourse is not a prerequisite of female orgasm; the pleasure provided by intercourse is to some extent independent of that which leads to clitoral orgasm.

The reader may recall that the central joke in *Deep Throat* is that Linda has a clitoris located in the back of her throat rather than in her vulva, the result being that she can only obtain satisfaction through fellatio. A frank and incisive comment from one of Hite's women is also illuminating. Asked to describe her concept of 'Sex in the best of all possible words' she replied 'My clitoris would be in my vagina, for Christ's sake, so I could come when I fuck.' A similar point is made in the ironic writings of Alex Shulman: 'Masters and Johnson observe that the clitoris is automatically stimulated in intercourse since the hood covering the clitoris is pulled over the clitoris with each thrust of the penis in the vagina – much I suppose as a penis is automatically stimulated by a man's underwear whenever he takes a step' (quoted in Hite, 1976). Altogether, there is insufficient reason to suppose that female genitalia are designed in such a way that heterosexual intercourse is the best way of achieving orgasm, and this being the case, it is difficult to credit that female orgasms have evolved in the human female in order to confer reproductive advantage.

Possible disadvantages of female orgasm

Symons' theory of the female orgasm seems best able to account for the facts as a whole, even though the other theories may each contain elements of truth. It seems that the vagina has been designed to produce pleasurable stimulation during intercourse, though not necessarily orgasm. There is even a possibility that orgasm may reduce a woman's chances of reproduction.

Masters and Johnson advise women who want to conceive not to have orgasm because, they say, the vasocongestion produced by sexual excitement constricts the outer part of the vagina and acts as a stopper to retain semen. Orgasm dissipates this vaginal plugging effect thus allowing semen to escape. However, at present it seems very doubtful that this mechanism has much influence on the likelihood of pregnancy, and in any case there have been suggestions to the contrary. Fox and associates (1970) studied pressure changes in the uterus during intercourse with the aid of a telemetric implant and have come to the conclusion that the resolution phase after orgasm is accompanied by a drop in pressure inside the uterus. This, they speculate, might serve to suck seminal fluid up through the cervix and so enhance the chances of pregnancy. Desmond Morris has proposed that orgasm may have evolved to keep the woman lying down after intercourse so that semen does not run out of her vagina before there has been every opportunity for conception. Neither of these proposals has any convincing empirical backing, so at the moment we can only conclude that there is no satisfactory evidence that female orgasm alters the chances of pregnancy one way or another.

A rather more likely disadvantage of female orgasm is that of premature termination of intercourse as suggested in the jackpot theory. If a woman reaches satisfaction before her male mate she might cut off proceedings before the biologically significant event. In addition, if women were identical to men in the ease with which they could obtain gratification from the sex act, this might interfere with their general reproductive strategy of choosing the best available father for their children and maximising the return for sexual favours bestowed.

One thing seems fairly clear. If the female orgasm performed any very important evolutionary function, natural selection would have ensured that it would occur with a great deal more ease and with much greater reliability than it does. As Symons notes, adaptations that have been arrived at by natural selection can normally be recognised by their precision, economy and efficiency. Evidence that orgasmic females enjoyed greater reproductive success than non-orgasmic females among our ancestors is totally lacking. In fact, if non-human animals don't have orgasm in the wild it is difficult to see how orgasm could be selected for.

It therefore seems that orgasm is best viewed as a

potential that all female mammals have. Humans differ from other species mainly in the extent to which this capacity has been developed, by techniques of foreplay, clitoral stimulation and extended intercourse. Nicholson (1979) has argued that women miss their orgasms because humans have chosen a bad position for intercourse, the 'missionary' position. This, he maintains, is the least likely to produce clitoral stimulation. The truth is nearer to the opposite. Human females are more likely to have orgasm than other animals and this may be partly because face-to-face intercourse produces more clitoral stimulation than 'doggy' fashion (entry from behind). It is true that the less common 'female superior' position (woman on top but still face-to-face) may be better still because it gives the woman more control and freedom of movement as well as better clitoral contact. This, however, is not a position that has been favoured either by our animal ancestors or the 'noble savages'.

Biological versus social functions

When female orgasm does occur it is virtually identical to that of the male because it employs the same neurology, built into the foetus before it starts to differentiate as male or female. It is no coincidence for example, that the initial orgasmic contractions are spaced at 0.8 seconds in both sexes. Also significant is the fact that testosterone injections increase a woman's capacity for orgasm, along with the size of her clitoris (Kane, Lipton and Ewing, 1969; Carney, Bancroft and Matthews, 1978).

Of course, once it has been discovered, the female orgasm does acquire *social* functions. Women who have had orgasms will want to have them again, men will want to give them to women (like a bunch of flowers), and a woman may want to have them (or fake them) to enhance her man's selfesteem and consolidate her bond with him.

But these are not the original reasons for the occurrence of female orgasms; they are merely social elaborations and uses that are made of it. Again returning to the analogy of piano playing, our fingers have not evolved for playing the piano, but if we once learn how to do it, it may have many social ramifications. We may give a great deal of pleasure to ourselves and to others. If we become very accomplished at the keyboard we may even earn a livelihood for ourselves and our families. But it would be patently foolish to argue this as the basis for the 'natural selection of piano playing ability'.

The conclusion to be derived from this discussion about the biological origins of the female orgasm is that its absence should hardly be considered pathological. It is not really a suitable case for treatment in the medical sense. Nevertheless, many women who do not have orgasms, or who have them less frequently than they would like, can learn to improve their chances and thus enhance their sex lives. Their male partners can also be educated to help. But part of this process of education will involve abandoning the myth that men and women are virtually identical animals. Women are not just repressed men; their sexuality is qualitatively different. I have no desire to 'deprive women of their orgasms' (an accusation that has sometimes been levelled at me); rather, I believe that a realistic appreciation of the facts of human sexuality will facilitate the 'therapeutic' process. Female orgasms do not come naturally whatever the social conditions; they always have to be worked at.

Multiple orgasms

One way in which female orgasmic capacity does differ from that of males is their celebrated ability, on occasions, to have multiple orgasms. Research on subjective descriptions of the experience of orgasm by men and women has shown this to be the only way in which orgasm gender can be distinguished (Vance and Wagner, 1976). Although it is not yet clear why this difference arises, it is likely that it has something to do with the female inability to ejaculate. Masters and Johnson have revealed that the neurological mechanisms underlying orgasm and ejaculation are to some extent separable, and it seems that it is the ejaculation which causes the male refractory period rather than the event of orgasm. According to Kinsey, prepubescent boys sometimes have the ability to have multiple orgasms, so women are more like boys than men in this respect. A Californian sexologist called Mina Robbins has been attempting to train mature men to recognize a distinction between orgasm and ejaculation so that they may be able to enjoy multiple orgasms, but this possibility remains very doubtful.

Male problems

The most common sexual problem complained of by men, or attributed to them by their partners, is that of arriving at orgasm too quickly. About half of all men report having experienced this difficulty at some stage during their sexual career, although of course the problem tends to recede as they get older and begin to get bored with the same partner.

In a sense, premature ejaculation is the flip-side of the female orgasm problem, since it is often defined as coming so quickly that the female partner seldom experiences orgasm. While this definition can be questioned on the ground that it requires that a rather phenomenal range of pre-ejaculation performance times be classified as 'premature', it does highlight the fact that quick orgasm in the man is just as much a female problem as a male one. Really, it amounts to the same issue that we have just been describing with the 'blame' reassigned from the female to the male partner. All of the same considerations therefore apply. Premature ejaculation is not really a pathological condition requiring 'medical' attention. Again, it is a natural product of the biologically determined difference in the speed of the sexual response cycles of men and women, and again there are techniques which can be learned by a couple which can improve their chances of mutual satisfaction.

Sex therapists usually prescribe the so-called 'squeeze technique' in which the woman gives her partner exercises in ejaculatory control, masturbating him until he is close to ejaculation and then gripping the end of his penis firmly until the risk of ejaculation subsides. If this is repeated many times it is apparently quite successful in overcoming the problem. A rather more straightforward and positive approach is to instruct the couple to repeat intercourse as soon as possible after the first (unsatisfactory) performance. Men are very seldom premature the second time around. Of course, they may not realize that they can do it at all, and therefore a certain amount of resourcefulness on the part of the female partner may be required. She may need to be taught that after orgasm a man cannot easily get rearoused all by himself; he may need some oral, manual or mental stimulation, but the woman will be a beneficiary in the end.

The other major sex problem experienced by men is inability to obtain an erection (impotence). This is rather more difficult to deal with since it may arise from a great variety of different causes, some of which *are* abnormal in the medical sense. A wide variety of physical diseases such as diabetes or brain lesions may be responsible for impotence, as well as temporary conditions such as tiredness or drunkenness. Testosterone injections may sometimes be helpful if the man's own hormone supply is insufficient for any reason, for example in the case of older or castrated men.

Of the non-physical causes, the one most commonly assumed to be involved is fear or anxiety, whether this is psychoanalytically interpreted in terms of the 'castration complex', or behaviourally, as a vicious spiral that has become established as a result of previous experiences of failure. The approach of most sex therapists, following Masters and Johnson, is to set up relaxed and secure conditions and remove performance fears by instructing the couple to engage in various kinds of foreplay without attempting full intercourse. The expectation is that an erection obtained under such an arrangement is unlikely to be wasted and that any experience of success will break through the malevolent spiral and throw it into reverse. Often this does seem to work.

In view of our discussion of the variety-seeking nature of men, however, it should be realized that impotence may also be due to boredom, anger, or some other resistance to the particular partner. This will be an obvious diagnosis if the man's impotence is specific to his wife or steady partner and his sexuality is known to be fully functional elsewhere. But some male patients may not have had the opportunity or courage to try out their erectile capacity in other directions, or their morality may rule out such a course of action. Sometimes, when the couple are seeing a sex therapist jointly, the husband may be unwilling to reveal that he is 'superstud' to other women. I know of at least one case in which a wife was earnestly seeking treatment for her husband's 'impotence', which was really due to satiation and exhaustion from his many extramarital affairs. When the Pope presented his recent dictum that 'A man who looks at his wife in lust has committed adultery in his heart,' one British radio raconteur observed that 'The man who doesn't has probably committed adultery in his lunch hour.' Relationship problems are notoriously difficult to treat, but some forms of impotence may be helped if the wife can permit herself to recognize her husband's need for novelty and arrange for some additional titillation either in fantasy or reality.

The conclusion that I want to stress is that sex therapy could be considerably sharpened and rendered more effective if the differing nature of male and female sexuality were fully recognized. Over recent years there has grown up a myth of natural compatibility between husband and wife, together with the idea that sexual difficulties are symptomatic of some kind of abnormality which may be treated within a medical framework. This idea, propagated in large part by the writings of Masters and Johnson, is grossly misleading. The majority of sexual difficulties are a direct result of biological differences between men and women which manifest themselves quite naturally. Men tend to arrive at orgasm more quickly and easily than women, and they have a greater lust for novelty and variety. For women, good sex usually requires security; for men, a degree of excitement is often necessary. There are many exceptions to these general rules, and a great deal of overlap between the sexes, but unless the average gender differences are recognized and taken account of men and women will continue to misunderstand each other and sex therapy will itself be fairly impotent.

7

Variant Sexuality

This chapter is concerned with those sexual behaviours that used to be called 'perversions'. A few decades ago the word 'deviation' became more widely used, and recently the euphemistic process has been taken one step further and the term 'sexual variation' is favoured. All of these terms mean much the same thing, although they differ in evaluative tone and some writers (such as Stoller, 1975) believe in retaining the word perversion to apply to sexual variations that contain strong elements of hostility toward unwilling human sex objects. Certainly, the different variations may be classified according to the extent to which innocent parties are harmed or humiliated, but they are all more or less frowned upon by large parts of the community. Apart from rape and sexual assault, which are unacceptable because they violate the rights of other people, most of these variations have in common the likelihood that sexual energies will be directed toward behaviour that could not theoretically result in reproduction. This seems to be their prime defining characteristic.

The outstanding fact about these variations, which is probably the key to understanding their origins, but which is curiously glossed over in most accounts of the topic, is that they are all very much more common in men than women. Fetishists, exhibitionists, voyeurs, frotteurs, paedophiles, zoophiles and sadomasochists are almost always men, and with most other variations, such as homosexuality, transvestism and transsexualism, men outnumber women to a very high degree. One explanation that has been offered for this fact is that of selective interest by legal authorities. Men are unlikely to complain to the police if a woman exposes her genitals or makes physical advances; unsolicited sexual initiative on the part of women may be regarded with surprise or suspicion by a man, but he is seldom upset to the extent that he seeks protection from the law. But this hardly accounts for the fact that all 'kinky' sexual activities are much more often pursued by men than by women, and that a large trade in pornography and prostitution has grown up in response to these interests. When women do engage in deviant sexuality it is frequently for non-sexual purposes, for financial gain, entry to higher circles, or to please a loved partner. True females deviates are few and far between.

The target-seeking nature of male sex drive

One of the main reasons why men are more likely than women to acquire unusual sexual attachments is the active and visual nature of their sex drive. Whereas female animals vary in receptivity according to the cyclic changes in their oestrogen levels, male hormones engender a tendency to scan the environment for suitable sexual targets. We have seen that men are more readily aroused by visual configurations, such as buttocks and breasts, than women are by any aspects of the male form. Apparently, men are also more susceptible to inappropriate visual fixations of the kind that we call fetishisms, whether we characterise the associative process as imprinting or conditioning. This is not because males have superior visual acuity to women; women can see perfectly well. The difference is to be found in the readiness with which connections are formed between visual stimuli and sexual arousal. Women are not turned on by static visual formations so much as by demonstrations of long-term interest and prowess on the part of male suitors.

This visual orientation of the human male is clearly evidenced in sexual fantasies men report. In my own survey and those of other researchers (Wilson, 1978) it emerges that, apart from group sex, the most frequently occurring theme of male fantasies could be described as visual or voyeuristic. This commonly involves reference to clothing such as black stockings and suspenders, high-heeled shoes, leather gear or nurse's uniforms, but other typically male fantasy elements that may be related to the visual emphasis include details of anatomy (such as colour of pubic hair or size of breasts), reference to the age and race of the partner, and descriptions of the sexual activity engaged in. Only very occasionally do women refer to the physical characteristics of their fantasy partners; the most common themes for women revolve around the identity of the partner, the amount of attention indicative of love and devotion that he is demonstrating, and the romantic, exotic and peaceful nature of the setting (islands, beaches, waterfalls, moonlight, candlelit dinners, fur rugs in front of an open fire). Clearly, men are primarily concerned with the physical form of their partners, while women are more preoccupied with the context and the emotional aspects of the relationship.

Innate releasing mechanisms and imprinting

Studies by ethologists such as Lorenz (1952), based on the observation of a variety of species from birds to primates, have led to the conclusion that sexual attachments are arrived at by two major instinctual mechanisms. Firstly, the range of potential sex objects is restricted to certain broad classes of stimuli by inborn neural circuits that are called *innate releasing mechanisms* (IRMs). Many species have an innate ability to recognize and respond sexually to adult members of the opposite sex of their own species. A male chimpanzee reared in total isolation from members of its own species is still sexually excited by the rear-side presentation signal of the female chimpanzee (although knowing what to do about it is another matter which does seem to require a certain amount of trial and error). Since this applies to our nearest relatives it almost certainly applies to us too, and we have previously discussed the likelihood that the visual configuration of paired, pink hemispheres, typified by female buttocks and breasts, is an innate sexual stimulus to the human male (Chapter 4).

The second instinctual mechanism is *imprinting*, which has also been discussed. At certain stages of maturation, particularly in early infancy, the range of stimuli which are to become sexually exciting in adulthood is further delimited and specified. Imprinting adds detail to the blueprint for arousal that was broadly sketched by the innate mechanisms, and to some extent it depends upon the visual stimuli that are available in the environment of the animal. Thus zookeepers are prone to become the sexual targets of a wide variety of animals in their care if these animals do not have sufficient exposure to members of their own species. Owners of pet cats and dogs are subject to the same effect, often to their embarrassment with neighbours and guests. The importance of the environment is also apparent when it is considered that rubber fetishism was not possible before the advent of rubber.

Either of these instinctual mechanisms can go wrong

(as can any other finely tuned piece of neurology) so that sexual responses can be attached to a class of stimuli that appears peculiar and socially unacceptable. Sometimes the categories of arousing stimuli are left too broad. One rather bizarre sex deviation that crops up occasionally is that of men who are sexually excited by other men and by certain animals (usually domestic) of either sex. They are not, however, sexually aroused by human females. The interest in animals, which may be observed as much in masturbation fantasies and wet dreams as in actual behaviour, cannot be explained as a form of substitute gratification, since those patients often have plenty of opportunity for contact with human males. Also, it cannot be explained away in terms of the 'village idiot' syndrome; among patients in this category described by Pinkava (1971) were a senior clerk, an antique expert, a surgeon, an organist and a university teacher. While one could resort to a Freudian explanation in terms of the 'castration complex' causing these men to be afraid of women, this seems less convincing when applied to similar phenomena in non-human animals. The most satisfactory theory is one of impairment in the innate releasing mechanisms which normally specify sex targets as members of the opposite sex of one's own species. The homosexual choice seems to imply an intact gender discrimination process within the 'home' species, but an inverted preference mechanism.

Other sexual deviations, again particularly in men, seem to involve an over-detailed specification of the sex object. The classic example is fetishism, in which some article or material becomes the focus for sexual arousal (Gosselin and Wilson, 1980). These attachments commence very early in life, often being recalled as well established by the age of four (which is about as early as we can remember anything). They are very resistant to any form of psychotherapy whether analytic or behavioural, and they often become more insistent in middle age, perhaps because a decline in general libido makes the individual more dependent on the fetish for arousal. Certain characteristics of popular fetish objects make the impairment of imprinting a very plausible explanation. They usually have strong gender associations and are worn close to sexually arousing parts of the body – as with high-heeled shoes, underwear, leather belts and buckles – and they often have striking visual attributes that are reminiscent of genital signals, being wet, shiny, black, pink or furry, for example (Epstein, 1975). Quite a high proportion of men find fetish materials erotically enhancing to some extent; what characterizes the full-blown fetishist is his tendency to prefer the sauce over the meat. His imprinting may be misdirected, but more obviously it is overspecified.

The fact that fetish targets usually appear in conjunction with biologically ideal targets (women and their genitals) is sometimes taken to support a learning or 'conditioning' theory of the acquisition of fetishisms. Indeed, it does seem possible to condition sexual arousal to pictures of boots by pairing them in a laboratory with pictures of nude women (Rachman and Hodgson, 1968). But this conditioning experiment does not provide an entirely satisfactory model for the acquisition of clinical fetishism, for a number of reasons. It does not explain why the true fetishist is not bothered about whether or not his fetish object is accompanied by a female in the flesh, or why the real woman ceases to be sexually arousing after the fetishism has become established. Also, it cannot easily explain why fetishisms are so resistant to extinction even though they become totally detached from the original source of arousal. Conditioned responses will normally disappear in the absence of reinforcement as did the fetishisms created in the laboratory by Rachman and

Hodgson. In these respects, the conditioning model of fetishisms appears unconvincing.

There is another explanation of the close association between fetish object and actual female which fits the faulty imprinting theory. Suppose that normal imprinting upon female genitalia requires that the infant male has some visual exposure to them. If a woman's private parts are never on display at the critical time for imprinting, the mechanism may seize upon the nearest visual configuration that is actually available. Women's underwear occupy the crotch area and often take on the shape of the pubis and vulva to some extent; so they are likely candidates for sexual imprinting. If pheromones are also involved in eliciting the imprinting process, pants would be at a particular advantage since they are steeped in vaginal odours. Shoes, especially black and high-heeled, also bear some similarity in size, colour and shape to the pubic triangle, so it is perhaps not surprising to find that they are also popular fetish objects. Belts, suspenders and fabrics such as fur and leather may become focuses for sexual arousal for similar reasons.

Note that this theory of the origins of fetishisms differs from both the conditioning model and the psychoanalytic theory. The conditioning model supposes that the fetishism is established by simultaneous (or near simultaneous) presentation of the fetish with a high level of sexual arousal that is originally a response to an actual woman or her genitals. By contrast, the imprinting model suggests that *unavailability* or partial masking of what, in conditioning theory, is called the 'unconditioned stimulus' is basic to the acquisition of the fetishism. Psychoanalytic theory emphasizes *symbolic* associations between the fetish and female genitalia. For example, a shoe is taken as equivalent to a 'substitute penis' which the fetishist attributes to women because he cannot brook the idea of their 'castration'. The imprinting theory assumes a direct *sensory* association which is basic to the choice of substitute sex target – shoes share certain visual, tactile or olfactory qualities with the female pubic area.

Two strands of evidence would seem to support the present conception of sexual deviation. One is the frequently reported finding that deviant men of all kinds tend to come from families which are sexually restrictive and which do not permit nudity in the house. Sexually deviant men are also less likely to have seen pornography as children (Goldstein *et al*, 1971). A popular interpretation of these facts is that early exposure to sexually explicit materials provides 'innoculation' against the potentially harmful effects of pornography later on in life. But a far simpler explanation is that the sight of female genitals in early childhood is a prerequisite for appropriate sexual imprinting and development. If this theory is correct we might expect to find fewer fetishists and sexual deviates in the future, since society as a whole is becoming more and more tolerant of nudity in and outside of the home. Certainly, my own research with Chris Gosselin indicates that the vast majority of men with fetishistic interests are currently well into middle age. There are, however, other possible explanations for this fact, such as the tendency mentioned above for fetishisms to augment with age.

The second kind of evidence that supports a 'faulty wiring' view of male deviation is directly neurological. If fetishisms and other compulsive sexual interests are due to impairment of subtle neural circuits laid down before birth and shortly afterwards, we would expect them to be associated with other indications of minor brain damage occurring in infancy. Indeed this is the case. An association of temporal lobe epilepsies with fetishism and transvestism in men was observed by Epstein (1961) and a similar connection with transsexualism has been reported by

Hoenig and Kenna, (1979). There have also been several clinical reports of compulsive sexual deviations that have been surgically removed along with a focal epilepsy. Research by a Czechoslovakian team (Kolarsky *et al*, 1967) showed that a wide range of male deviations including homosexuality, sadomasochism and exhibitionism as well as fetishism, were associated with minor damage to the temporal lobe, and what is most interesting, this damage only affected sexual preferences if it occurred before the age of three. Typical brain lesions associated with sexual deviation in adulthood were those due to head trauma during birth (often associated with forceps delivery) and meningitis or encephalitis contracted within the first year of life. The brain damage involved was usually very subtle and could only be detected by EEG examination or clinical symptoms relating to epilepsy. Most of the subjects were of normal intelligence and free of gross psychiatric disturbance. This raises the question of whether other indications of unusual or impaired brain wiring would also be connected with deviant male sexuality – symptoms such as stammering, tics, aphasias, amnesias, and mixed dominance of the left and right hemispheres of the brain. In this connection it is interesting that the Yorkshire Ripper was recently thought to be compensating for a stammer in the recording that he sent to the police. While some psychoanalysts such as Otto Rank have talked of 'birth trauma' being involved in the origins of neurosis, they were apparently talking about psychological stresses imposed by the shock of being born rather than minor brain damage occurring at that time.

Evidence for neurological involvement in sexual deviations leads to further speculation about the reasons for the male preponderance of this kind of behaviour. Male brains differ on average from female brains in the degree of specialization of the right hemisphere – the side of the brain that deals with spatial configurations. The importance of the right hemisphere in males may well have something to do with their visual sex orientation, while the left-hemispheric advantages of women (such as their superior verbal processing, sequential thinking and future time perspective) would seem well-suited to the support of their particular mating strategies. Now minor damage such as might occur in forceps delivery or infantile meningitis may be less critical to the female brain because the second hemisphere acts as a back-up system. In the case of the male brain, the second hemisphere does not replicate function to the same extent and so damage to one temporal lobe is more likely to be manifested in peculiar behaviour. A parallel may be seen in the specialization of the Y chromosome in males which leaves men more susceptible to a wide variety of sexlinked recessive disorders such as colour blindness and haemophilia. This idea of male vulnerability arising out of hemispheric specialization is highly speculative at present but certainly merits further consideration as our knowledge of neurology advances.

Embryonic development

Since all foetuses develop as female unless instructions to the contrary are received via the hormones, there is a sense in which male brains undergo a greater conversion process during embryonic development. Most of the research indicates that it is the last third of pregnancy that is the critical period for this gender differentiation of the mammalian brain. Now since it is the male brain that needs to be changed from the standard pattern of development, it follows that there is a greater chance that something can go wrong with the brain mechanisms involved in sexual choice. The male foetus has developed male genitalia well before this brain conversion takes place; in fact, it is the male gonads of the foetus that release hormones which modify brain development in the male direction. If the parts of the brain that are supposed to receive the masculinization message from hormones in the blood stream fail to receive that message for some reason, they will remain female components in an otherwise male body. For several years it has been supposed that this might be an explanation of some forms of male homosexuality and transsexualism.

Such a theory is consistent with the fact that male homosexuals and transsexuals outnumber their female equivalents by a considerable factor. One could conceive of the reverse happening – male hormones somehow finding their way to the critical hypothalamic nuclei in a biological female – and there is some evidence that this can happen if the mother of a girl foetus is treated with male-type hormones in the latter part of her pregnancy (Reinisch, 1977). But of course such an event is bound to be relatively rare and therefore it is not surprising that we observe fewer disorders of sex orientation in females.

This prenatal hormone theory would also explain why sexual identity disturbances in boys who will grow up to be homosexual or transsexual in adulthood can be observed from the earliest years (Zuger, 1978). Equally, it explains why these individuals can not necessarily be discriminated from normal males in terms of the amount of male hormone circulating in adulthood; if certain switches in the brain were not masculinized before birth they will not respond to circulating testosterone after puberty. Overproduction of male hormones has been noted in a high proportion of female transsexuals (Sipova and Starka, 1977) but this might also be attributable to events occurring in the critical prenatal period. Male hormones given to a normal adult female do not alter sex orientation however much they may enhance libido or masculinize the body.

Male competition

Another reason why men are more likely to adopt deviant sexual behaviour than women is the intensity of inter-male competition for mates. We have seen that in most mammalian species the more successful (dominant) males monopolize more than their fair share of females with the inevitable result that some males must miss out altogether. This leaves a great deal of male libido unsatisfied and so there is a great need for alternative 'outlets'. Masturbation is one obvious outlet, but it is lacking in important mental and emotional components of the complete sex act, and so a great many other substitutes are resorted to. These include pornography, rubber blow-ups of women purchased from sex shops, paedophilia, animal contacts and homosexual encounters. A large proportion of the homosexual behaviour that occurs in prison may be accounted for in this way. It is not 'true' homosexuality because women remain the ideal sex targets, featuring in fantasies and being reverted to immediately upon release from prison. The same pattern also occurs in all-male boarding schools. In this context also, males usually become substitute sex targets rather than actual preferences.

It is likely that even outside prisons and public schools some men feel unable to compete for women, and find it a great deal easier to pursue sexual outlets in the gay community. Perhaps they have had unfortunate early encounters with women which led them to feel incompetent or unattractive. Perhaps they were savagely punished for heterosexual play. Perhaps they hated their mother and generalized this attitude to women at large. Perhaps they are afraid of pregnancy or commitment. For whatever reason, these men decide to opt out of the heterosexual rat race and take their pleasure with their own kind.

From the point of view of survival of the species, it does not matter a great deal if a proportion of males are removed from the effective breeding pool by their pursuit of non-reproductive outlets for their sex drive. All the females can be (and usually are) fertilized by the remaining males. By contrast, if women were to adopt deviant sexual practices that did not lead to impregnation, there would be an irrecoverable loss of reproductive efficiency to the species. Perhaps it is partly for this reason that this does not happen to any great extent. Most women, however unattractive relative to their peers, are able to persuade some man to service them without too much reluctance. The result is that, as with other animal species, human females breed fairly evenly while males are highly variable in terms of the number of offspring that they sire.

Exploratory deviation

Since males tend, on average, to have high libidos, their sexual interest is inclined to generalize quite widely beyond the confines of their ideal target. A certain amount of sexual behaviour that would be classified as deviant may therefore be attributed to 'overspill' rather than substitution. Many husbands who have a good, active sex life with their wife report that they occasionally also enjoy masturbation, just for a change. Likewise, some men report having been motivated to try homosexuality or bestiality not because they expected to prefer these activities over heterosexual intercourse as a steady sexual diet, but purely in a spirit of exploration. Consistent with this 'overspill' theory is the finding the bisexual men show indications of higher libido than either exclusive homosexuals or exclusive heterosexuals (Wilson and Fulford, 1979). There is a large body of evidence in psychology to support the idea that the stronger a state of drive or deprivation becomes the greater the range of stimuli that are capable of evoking behaviour that will lead to its being consummated; apparently this applies just as well in the sexual sphere. Again, there is theoretical reason to suppose that men will be more likely to adopt deviant behaviour than women; men are presumed to have a chronically stronger sex drive. It is also questionable whether exploratory sexual variations should the classified as deviant at all, since the primary, preferred sexual target remains 'normal'.

Genetics and homosexuality

To some extent the predisposition toward homosexuality appears to be genetically determined. Kallmann (1952) investigated thirty-seven pairs of identical twins in which one member of the pair was homosexual, and found that in every case the other member of the pair was also homosexual. By contrast, a sample of non-identical twins showed only 15 per cent concordance for homosexuality. Heston and Shields (1968) reported on a rather interesting family of fourteen children which included three sets of identical twins. Two pairs of twins were concordant for homosexuality, while both members of the third pair were heterosexual. Since the time of Kallmann's study, several pairs of identical twins have been reported upon who were not concordant for homosexuality, so the genes do not provide a total explanation. Nevertheless, it does appear that genetic factors are involved in the transmission of homosexuality.

From the evolutionary point of view this is something of a puzzle. Surely any gene predisposing towards homosexual preferences would quickly eliminate itself because homosexuals produce fewer children than heterosexuals. Yet the survey evidence suggests that there is a reasonably high proportion of people in any time or culture (perhaps four or five per cent of men and one per cent of women) who are exclusively homosexual throughout their lifetime, and there is no sign of this proportion diminishing from one generation to the next.

A number of explanations of this paradox have been offered. One is that the predisposition to homosexuality may be spontaneously supplied as a fairly common mutation that has reached a kind of steady state. That is, the creation of new homosexuals by mutation may be exactly equal to the rate at which they are extinguished by reproductive deficiency. Such a state of affairs could also apply to a number of other genetic anomalies that are clearly disadvantageous, such as haemophilia – a deficiency in the blood-clotting facility which can lead to fatal bleeding. Haemophilia, like colour blindness, is a sex-linked recessive characteristic, which means that it does not manifest itself in all individuals who carry the genes, and it hardly ever appears in females, since an anomaly on one X chromosome is likely to be overridden by the 'healthy' information on the other. Although it does not appear to follow simple Mendelian laws of inheritance, a genetic tendency to homosexuality could survive more readily as a recessive than a dominant trait, and sex-linkage might help to explain the male preponderance. In any case, susceptibility to a wide variety of diseases, including schizophrenia, cancer and heart disease, is known to be influenced by genetic factors, so there is no real reason to suppose that anything genetic must necessarily be advantageous.

Hutchinson (1959) suggests that homosexual genes may result in some reproductive advantage when a person has genes for homosexuality that are dominated by genes for heterosexuality. Such a person may be a superior breeder to the one who possesses only genes for heterosexuality. This possibility cannot be ruled out, and the Wilson and Fulford finding to the effect that bisexual men display greater sex drive and activity than exclusively homosexual or heterosexual men might perhaps connect with it. Perhaps the degree of discrimination with respect to sex objects is a mediating factor that accounts for the survival of homosexual genes; high generalization of the male sex drive would take in a greater variety of women, including those that are less attractive, as well as extending to male targets. Such a theory does not, however, explain the appearance of exclusive homosexuality.

E.O. Wilson (1975) suggests that homosexual members of primitive societies may have been particularly useful to their tribe because they could adopt a third, independent sex role. Being free of the responsibilities either for hunting or child rearing, they could be particularly useful in providing assistance for either the men on the hunt or the women in the domestic situation, and this would help the group as a whole to survive. This theory implies kin selection – the idea that behaviour may be selected if it is sufficiently helpful to the group that carries many of the same genes even though it is detrimental to the individual concerned. Although this clearly does apply to insect societies, its relevance to primates is less certain.

What seems to me a more likely hypothesis is one that admits no genes predisposing specifically to homosexuality. The genetic contribution to homosexual orientation could be more of a negative one, that is, an absence of sufficient masculinity to make one successful in male competition. The males who are successful in competing then monopolize an unequal share of the female resources, have more offspring, and their strengths are emphasized in the following generation. Of the males who miss out, some might find it comfortable and convenient to assume certain aspects of the female role, being submissive in relation to the dominant males, thus appeasing them and sharing in the protection that they offer to the females. Others, feeling a need for some kind of substitute sexual activity, might find it easier to make contact with these submissive males than compete against the top men for the females. We have already noted the tendency of men to adopt homosexual behaviour as a substitute for heterosexuality in the forced absence of women, as in prison or public schools.

This implies that the biological basis of homosexuality, such as it is, is much the same as the biological basis of dominance versus submission. Those males who can successfully compete for women become heterosexual; those who are less successful adopt one of several possible alternatives which we label as sexual deviations, one of the most common of these being homosexuality. If we are to identify any benefit to the social group as a whole, it is that reproduction is left to those males with the most vigour and strength to contribute to the species.

Since this sounds rather insulting to homosexual males, it is fair to redress the balance by pointing out that, whatever the causes of their anomalous sex orientation, homosexuals have proved to be of great benefit to society throughout history, particularly in the arts. Having some of the male qualities of persistence and creativity, but without the responsibilities assigned to either partner in the reproductive relationship (breadwinner or nurse), homosexual men have contributed greatly to the culture as philosophers, writers, artists and musicians. To what extent this is vital to the survival of our society, as implied by the kinship selection theory, is open to question, but gay men (and women) have certainly improved the quality of life for all of us.

Support for the idea that submissiveness is basic to male homosexuality comes from the observation of monkey behaviour. If a male monkey is sufficiently cowed by another male he will sometimes adopt a female gesture as a form of defensive appeasement. The most explicit of such gestures is that of bending over and presenting the backside to the other male in the manner of the female invitation to copulate. Unable to compete successfully as a male, some monkeys will thus, temporarily at least, adopt the female role, perhaps until a better opportunity to assert their masculinity arises. I once witnessed an exchange between two male motorists in a busy London street which was amusing if not necessarily relevant. One motorist got out of his car to remonstrate with the driver in front who was stopped at a red light. The driver still seated in his car wound down his window and delivered two or three well-timed blows into the face of the complainant. At this point, the driver standing on the road, still smarting from the blows, lowered his trousers and turned to present his bare backside to the driver inside the car. This gesture was no doubt primarily intended to be satirical and insulting, but one cannot help wondering if it did not also relate in some way to homosexual submission. Certainly, it put an end to the fight and the two motorists resumed their journeys.

Male versus female homosexuality

According to the feminist point of view, men and women are much the same by nature but have been led by sex role learning to assume widely discrepant life styles. The opposite (evolutionary) hypothesis is that men and women are basically very different animals but the difference between them has been masked or minimized by the requirements of social living. Since the ideal male and female strategies are in direct conflict, the two sexes have to meet half-way in order to establish any communication or working arrangement. Homosexual people are therefore of particular interest to the evolutionary theorist because they provide an opportunity to look at the way in which men and women behave sexually when they have no need to compromise with the differing proclivities of the opposite sex. Furthermore, homosexuals can hardly be said to be behaving in accordance with stereotypes or social expectations since society generally deplores, or at least derides, their behaviour, which tends to be almost directly counter to social expectation. They may learn certain tricks or conventions within the homosexual subculture, but it is most unlikely that parental treatment or the influence of other social agencies could be of major significance since the behaviour is so deviant in itself.

If social learning of sex roles is of major importance, men and women should be relatively undifferentiated when it comes to homosexual contacts. In fact, the differences between men and women seem to be magnified in homosexual relationships. Men tend to be promiscuous and impersonal in their encounters with other men to a quite remarkable extent. According to one estimate (Bell, 1974) they average over a thousand partners in a lifetime, most of these being fleeting contacts made at gay bars and parties, parks and public toilets. This compares with an average of about ten different female partners for the heterosexual man. While long-term affectionate relationships ('marriages') between men are not unknown, they are much less common than brief encounters. Most homosexual men have no desire to get to know their contacts before having sex, or ever to see them again afterwards.

Lesbians behave quite differently from homosexual men. They tend to have fewer relationships, perhaps only two or three in all (which is less than heterosexual women), and these are intimate and caring – much like the relationships that heterosexual women seek with men. More often than not, lesbian sexual involvement grows out of a deep and longstanding friendship with the partner and is seen by both partners as a natural outgrowth of that relationship.

In a survey of men and women who engaged in bisexual behaviour, Blumstein and Schwartz (1976) found similar gender differences. Men were more likely to have their first homosexual contacts with strangers or male prostitutes whom they would probably never see again. They were apparently also more likely to conduct homosexual and heterosexual affairs in parallel. Women more often had sex with a close female friend, seeing this as a development of their strong emotional attachment, and were likely to alternate their orientation rather than engage in simultaneous affairs with men and women.

The interpretation that bisexuals put on their own behaviour is also illuminating. Many men viewed their homosexual contacts as a kind of modified masturbation, a substitute diversion engaged in primarily because females are difficult to obtain. One of Blumstein and Schwartz's subjects explained it thus:

> I'm straight, but I need outlets when I'm away from home and times like that. And it's easier to get them with men than with women. So I go into the park, or at a rest station on the highway and get a man to blow me. I would never stay the night with one of them, or get to know them. It's just a release. It's not like sex with my wife. It's just a way to get what you need without making it a big deal.

By contrast women usually saw their bisexuality as an extension of their natural female affectionate behaviour. Sex very often followed a period of deepening friendship and love with a person who happened to be of the same gender.

If the bisexual experience first occurred in the context of group sex or a threesome, the women found it a great deal less threatening than did the men – even though these events were usually inspired and initiated by the men. Blumstein and Schwartz thought this was because male homosexuality is more deplored by our culture than lesbianism, so that the men were more insecure in their sexual identity. More likely, the men found the presence of other males upsetting because their natural orientation to another male is competitive and aggressive, while women, presumably because of the requirement of the mother role, have the capacity to be loving towards people of either sex.

Gender differences are also apparent in the homosexuality that develops in prisons. Whereas with men the homosexuality serves purposes either of sexual release or of dominance, homosexual liaisons in women's prisons seem to be connected with the development of a quasi-kinship system – the women variously adopting the roles of wife, husband, aunt and so on (Giallombardo, 1974).

Clearly, male and female homosexuality is very different in character and constitutes a fairly impressive test case for the evolutionary theory of gender differentiation. In a way, it is surprising that such startling differences should appear since we have already suggested that in a certain proportion of homosexual men there is a kind of reversal of brain mechanisms that makes them effectively female as regards sex orientation. Apparently, this kind of reversal accounts for only a small proportion of male homosexuality, or else it is limited to the brain processes determining the gender of the sex target and not to other aspects of sexual behaviour, for it is clear that the majority of homosexuals do not adopt all aspects of the opposite sex role. Male homosexuals usually remain distinctly male as regards the exploratory nature of their libido and their concern with the youth and physical attributes of their partners. Lesbians retain their feminine concern with the emotional aspects of their relationships and their long-term prospects.

Sadomasochism

The practice of gaining sexual arousal from inflicting or receiving mild degrees of pain is perhaps the most popular sexual variation after homosexuality. At first sight this also seems extraordinary from the evolutionary point of view because pain is the guideline by which we avoid stimuli that are dangerous to us. Quite how this could come to be connected with the very positive pleasures of sex is something of a mystery.

Once again, an examination of the behaviour of our animal ancestors may help to throw some light on this paradox. Prior to copulation, many species (particularly mammals) engage in a kind of fight or struggle. Often the male will bite the female on the back of the neck in order to maintain a good hold on her while performing the sex act, and this no doubt inflicts a degree of pain. In some species this pain is known to be instrumental in releasing ovulation, and the fertility of the female is thus increased. This effect is presumably mediated by her emotions and endocrine system, so it is not unreasonable to suppose that the rough treatment she receives from the male is in some sense a 'turn on'. An evolutionary advantage can also be seen in the fact that the male has to be very assertive in order to pass on his genetic material, not just in competition with other males but also in gaining the cooperation of the female. By mating with males who pursue them with vigour and strength, the females are ensuring similar survival-related characteristics in their offspring.

Does this pattern of sex combined with brutality extend into human psychology at all? It has often been suggested that women invite and enjoy rape, usually most unfairly. Women don't usually want to be raped by the kind of man that commits actual criminal rapes which result in court convictions. But they do frequently fantasize about being taken forcefully by an extremely dominant man (being 'swept off their feet' as the saying goes) and there is little doubt that they enjoy male assertiveness and initiative (Wilson, 1978). The feminists who advocate female assertiveness in the bedroom have not succeeded in altering the behaviour of the average female to any great extent, despite the fact that most men say they enjoy women who are sexually aggressive. Therefore, some degree of masochism seems inherent in the female psyche.

There is also some evidence which could be interpreted as indicating increased fertility in women taken by force. Apparently, women who have been raped are much more likely to get pregnant than would be expected on the basis of a comparison with a similar number of consenting acts of intercourse (Parkes, 1976). This could mean that the special emotional circumstances surrounding the act of rape can somehow precipitate ovulation and thus increase the chance of pregnancy. There are, however, other possibilities. Perhaps men who rape girls are particularly potent and fertile at the time of the offence because they are in a state of deprivation. Perhaps rapists are generally more fertile because of high testosterone levels. Alternatively, some of the girls may have classified the origins of their pregnancy as rape because it was easier to offer such a story to their dismayed parents, husbands or boy friends. All told, it would be premature at present to conclude that rape promotes fertility.

If animal behaviours have any relation to adult sadomasochistic behaviour we would expect that masochism would be primarily a female predilection while sadism would more often be found in males. This appears to be the case. A general tendency for female fantasies to be passive and masochistic and male fantasies to be active and sadistic was found by Wilson (1978). When it comes to actual behaviour the difference is also apparent. Morton

Hunt (1975) asked a representative sample in the USA whether they had ever obtained sexual pleasure from inflicting or receiving pain. Although the actual numbers admitting to such practices were quite small (less than five per cent of the population), twice as many males as females reported sadistic pleasure, while twice as many women as men were masochistic. Given that replies to the questionnaire were anonymous and that the admission of such perverse sexuality is fairly unacceptable socially, it is most unlikely that social pressures are responsible for the gender differences in these reports.

Hunt's results imply that sadomasochistic deviations are, on balance, just as common in females as males, yet folklore and clinical impression suggest that most sadomasochists are male. Certainly, the vast majority of people joining clubs which cater to this preference, and buying pornography and equipment connected with it, are male. The few females studied by Gosselin and Wilson (1980) who did participate in such practices were apparently turning it on for the benefit of male contacts – either their lovers or professional clients.

How can these conflicting pictures of the sex balance be reconciled? Probably it is a question of how the sexuality comes to be classified as deviant. Most of the women who reported pleasure in receiving pain in the Hunt study, may have been referring to acts taking place within the context of a loving, intimate relationship with a particular husband or boy friend. These acts would include the receipt of playful spankings or being tied to the bed; they would also include normal intercourse taken to a painful extreme, for example, forced entry with insufficient lubrication or extended intercourse to the point of soreness. The male sadomasochism that is regarded as deviant, on the other hand, is more likely to be impersonal and involve torture equipment that is peripheral to normal sexuality, such as whips, studded and sharp objects, pincers, tongs, heated objects, or electrical equipment. Very often, it will have connections with homosexuality, fetishism, transvestism, bondage, or other such characteristically male variations (Spengler, 1977). Most of the men interested in such activities alternate the roles of sadist and masochist, to experience both ends of the stick so to speak, although they generally prefer the masochistic experience. It is probably these facts that cause such men to be thought of as deviant. In other words, it is the impersonality and role reversal, more than the predilection for pain, that is left to be explained.

Here again, I am inclined to favour an explanation in terms of the failure of some men to succeed in competition with other men for access to women. This leaves them with an adaptation problem. One solution to this problem is to engage in fantasy violence with other men in which they gain power and ascendancy, for example, chaining them to a wall and burning them with cigarettes. Or they may skip the struggle with other men and go straight to the fantasies of super-masculine power over women – hence the bondage, enslavement, spanking, etc. Again, and perhaps more commonly, they undergo a kind of role reversal – becoming submissive slaves, babies with dummies, or honorary females (transvestites). Such role-playing by-passes the need to exert dominance at all, and for such men sex as a victim may become just as exciting and satisfying as sex as a victor.

Concluding comment

These variant sexual behaviours, which are the speciality of men, may still appear as abnormal to the clinical psychologist. Insofar as we have implicated faulty neural circuits and hormonal processes they are perhaps legitimately regarded as pathological. And certainly, if the patient himself is unhappy with his condition and wants to change, psychological treatment may quite reasonably be attempted (indeed it may be unethical to withhold it). But what has been said in this chapter may help to put the phenomena of sex deviations into a broader perspective. Viewed within an evolutionary framework, many of the sexual predilections that have hitherto appeared as totally irrational and inexplicable may seem instead to be inevitable and almost natural. Perhaps if people better understood the origins of these practices they would be less intolerant of them. After all, they may sometimes be beneficial to society at large because of their eugenic consequences. It will also be clear that simple-minded attempts to eliminate sexual variations by conditioning methods or psychoanalysis are likely to be ineffective or even counterproductive. The biological origins and evolutionary significance of these behaviours, as well as the range of alternatives that are available to the individual, should be taken into account when attempts are made to 'treat' them. Little will be gained by imposing legal strictures upon mild sexual deviations, or applying medical 'cures', unless suitable substitutes such as prostitution and pornography are freely available in society or the social skills required for heterosexual seduction are properly taught.

8

Social Influences

Because in this book I have focussed on biological explanations of human sexual behaviour, particularly the light that is thrown upon it by evolution theory, the criticism that is bound to be made is that I have ignored cultural factors. Since it is not my intention to deny the importance of social learning, this last chapter will be devoted to the role of such processes in modifying the behaviour that arises from our natural dispositions. At the same time, I will consider the social implications of these evolutionary theories.

There is little doubt that we can control our biological urges within certain limits. The scientific question is to what extent we are *able* to do so, and the moral one is to what extent we *should*. It is important to separate scientific from moral questions because confusion between them lies at the root of many unnecessary arguments.

Kissing

I shall begin with a brief analysis of the subject of kissing, partly because it is an interesting topic that has not so far been dealt with, and also because it illustrates the complex mixture of social and biological factors that often enters into real-life sexual and romantic behaviour. Kissing is, in fact, an unusual form of human behaviour that has puzzled many observers. No single and simple biological explanation will do, for kissing varies in style from one culture to another. In our culture, lip-to-lip contact is a normal step on the scale of increasing intimacy between holding hands and outright intercourse, and one that is very seldom bypassed in respectable courtship. Eibl-Eibesfeldt (1971) suggests that this may have a non-sexual origin, deriving from parental feeding. In some primitive tribes (such as the Papuans) mothers feed their babies with pre-masticated food which is passed from mouth to mouth. This also used to be customary in some rural areas of central Europe. Even though this style of infant feeding is no longer necessary, lip contact could survive as a ritual gesture of nurturance and love. But this would hardly account for the Polynesian style of kissing which consists of touching cheeks or rubbing noses.

Probably the real key to understanding kissing is the extreme intimacy of the act. A degree of trust is involved in contact with the face, and particularly saliva, which can easily result in the transmission of disease. In this respect, the act of kissing may be an intimate act equivalent to sharing a bath or toothbrush. Truly, it may be just as well to let each other's microbes mingle and establish some kind of rapport before undertaking the considerable responsibilities of reproduction and child-rearing.

Kissing is also a kind of test. In making contact with another person's mouth or nose, it is inevitable that we inhale their bodily odours to some extent. If all goes well this may be a turn-on, but if halitosis is present (often a sign of ill-health) then it is as well to discover this early on in the courtship sequence. At a later stage in the relationship, when sexual activity has begun, oral sex may serve a similar function. Genital smells vary considerably according to health and the state of a woman's cycle.

The lips have a particularly heavy nerve supply, making them very sensitive, as well as putting them under very flexible muscle control. They need both these attributes as guardians of the oral portal. This sensitivity lends itself to the reception of various social and emotional signals from the other person, especially their receptiveness and sexual excitement. If both parties are sexually aroused, the reciprocation will lead to a kind of spiral of excitement which will promote sexual activity. If one party is unreceptive, this would normally be registered by the other who would be likely to withdraw, at least on that occasion. Thus the testing of sexual interest is probably a major reason for the popularity of mouth kissing in our culture. A girl may be obliged to accept a kiss at the end of a date, but she can communicate a great deal of apathy or even disgust in the process, if she is so inclined. Altogether, kissing is a much more direct and informative technique for assessing future sexual potential than verbal negotiation.

Finally, it is hard to escape the symbolic significance of some forms of kissing. The most obvious case is that of 'French kissing', in which the tongue is inserted inside the mouth of the partner. Psychoanalysts, as well as Desmond Morris, have pointed out that the mouth is an erogenous cavity with mucous membranes, and thus has a fairly clear association with the vagina, while the tongue is unmistakably phallic. When a man 'French kisses' a woman, it is probably part of his testing schedule. When a woman takes the initiative and inserts her tongue inside the mouth of a man, he is bound to interpret it as a definite sign of 'heat'.

Clearly, kissing is not without biological significance, but the social purposes for which it is employed are at least equally important to its understanding. To consider it only as a residue of mouth-to-mouth feeding or as a technique for erotically stimulating a partner would be to overlook some of its more important functions as an act of communication. The same applies to most other forms of sexual behaviour such as extramarital intercourse, homosexuality and sadomasochism. I have rather concentrated on their evolutionary significance at the expense of describing their cultural adaptations, but they could equally well be considered from the point of view of their many possible social meanings. Neither aspect can, by itself, provide a complete account of human sexuality.

Extremes of social intervention

Proponents of the belief that human sexual behaviour is largely culturally determined often point to the enormous differences that may be observed from one culture to another. Our own society is fairly free about sex. Premarital and extramarital sex hardly raise an eyebrow any more, and most sexual variations (with the exception of rape and paedophilia) are regarded more with amusement than horror. Some Asian and Pacific societies appear to be even more liberal than ourselves in certain respects, though it would be a mistake to imagine that they have no rules restricting sexual behaviour.

At the other extreme are societies which impose very tight controls upon the avenues of sexual pleasure and very severe penalties for transgression. W.S. Gilbert, even from a Victorian perspective, was aware of oriental sex repression when he wrote these lines:

> Our great Mikado, virtuous man,
> When he to rule our land began,
>   Resolved to try
>   A plan whereby
> Young men might best be steadied.
> So he decreed in words succinct,
> That all who flirted, leered or winked

(Unless connubially linked)
Should forthwith be beheaded.
(*The Mikado*, 1885)

We may think this idea very funny, yet some Eastern societies are perfectly capable of such brutality. In 1978 reports reached the United Nations in Geneva that the Khmer Rouge regime in Cambodia had executed at least twenty young men and women for flirting. Sex before marriage was subject to the death penalty, and according to some refugees, men were not permitted to talk to women about anything except business. It is understandable that in circumstances like this the rate of extramarital intercourse might fall dramatically and untutored observers might gain the impression that Cambodians were not very interested in sex.

Communist China has also treated the sex instinct as counter-revolutionary and sought to suppress it as far as possible. An official pamphlet first published in 1963, and still issued to young people today, makes it clear that nobody is supposed to have intercourse before the age of twenty-five, and that premarital sex is regarded as criminal. Masturbation is no solution to youthful frustration, because it may cause 'nervous weakness'. The establishment of 'correct Communist attitudes' is the best way to deal with the urge to masturbate, but it is also recommended to 'avoid all sexually-exciting substances such as tobacco and alcohol', shun pornographic books and films, and wear baggy underpants. As regards the frequency of marital intercourse, 'for the first few months after marriage, most couples will establish a routine of having sex once or twice a week. With increasing age and the gradual cooling of sexual ardour, the frequency of sex will drop, mostly to around once every one or two weeks.' These norms are very low by European standards and have apparently been set that way so that little time and energy is diverted from working for the State.

Arab cultures are remarkable for the extreme repression that is exercised against women in particular. In Saudi Arabia, women can never really go out alone, and they must be totally covered from face to ankle or they risk assault by the dreaded *mutawwa* (religious police). No mixed-sex activity that involves even the slightest degree of intimate contact is permissible. This makes it virtually impossible for women to go swimming or have their hair styled. While they may be permitted to work as nurses and teachers, girl students can have contact with male professors only through closed circuit television (*Daily Telegraph*, 4 September 1979). Adultery is rated as a most serious crime (at least equivalent to murder) and frequently results in public execution, particularly for female offenders. The cruel operations of cliroridectomy and infibulation which are still performed in many parts of North Africa may be construed as further attempts to curtail female sexual freedom and pleasure. With access to women so difficult, it is not surprising that Arab men have a reputation for buggery in their own countries and for patronising prostitutes in Europe.

Our own culture has seen remarkable changes over the last hundred years. In Victorian days, sex was not mentioned in polite company and nudity was strictly forbidden. Women were required to wear ankle-length dresses because the legs were regarded as too erotically stimulating to be exposed. This obsession reached ridiculous proportions with the institution of 'bathing machines' at the seaside and the (possibly fabled, or at least exaggerated) covering of the legs on grand pianos at public concerts. The Victorians were not totally asexual; brothels flourished and corporal punishment of children and servants was conducted with distinct sexual overtones. Nevertheless, the evident puritanism of Victorian times is usually seen as a reaction against the more playful previous era.

What does all this variability between and within cultures tell us? One thing it certainly does *not* mean is that people living within a repressive cultural climate are devoid of sexual instincts. The instincts remain much the same; only their manifestation is affected. Thus, if heterosexual outlets are made difficult to obtain, as in Arab countries, a certain amount of extra homosexuality is likely to occur (just as it does in prisons and public schools). If all outlets are suppressed, people may simply become frustrated and neurotic. This was one of the insights that made Sigmund Freud famous, and indeed he may have hastened the end of the Victorian era of sex repression by making people aware that others were just as sexually preoccupied as themselves and that driving the sex instinct underground can give rise to unpleasant side-effects.

One might ask why so many societies find it necessary to impose such severe restrictions and sanctions upon sexual behaviour. The reason is that sex is a potentially disruptive force. It is a very powerful instinct involving scarce commodities (attractive women) and can therefore provoke lethal conflicts in the same way as competition for other forms of power and property. Sexual mores are designed to minimize the risk that libidinous urges will lead to violence. If rights of sexual access are clearly defined (through kinship and marriage) and the temptation to transgress the social code is countered by the threat of severe punishment, social stability, it is hoped, can be maintained. Unfortunately, people tend to fall in and out of love inappropriately, and their lust is often too powerful to be contained. Societies therefore go through cycles of first trying to satisfy their own need for order and stability by imposing rigid rules of conduct, then relaxing the rules to allow for individual difficulties in conformity. Needless to say, there is no simple solution to the problem of reconciling social and individual needs.

Even if there is no question of conflict arising out of sexual competition and jealousy, many societies are concerned to restrict sexual pleasure because they see it as antithetical to work output. (For the same reason they are opposed to other hedonistic pursuits such as pop music and drug taking.) This is particularly true of political systems that we would characterize as totalitarian, and societies that are struggling to survive economically, like China and Cambodia. The greatest amount of sexual permissiveness is seen in countries which enjoy general political freedom and concern for individual rights and a comfortable existence as regards food, resources and climate. The United States, Western Europe and Polynesia are the areas that come to mind as fitting this description.

Within European society there is a tendency for sexual freedom to fluctuate in accordance with economic conditions. Morris (1977) notes that, over the last sixty years at least, the length of women's skirts has comprised a near-perfect economic barometer. As the stock market rose during the 'Roaring Twenties' and 'Swinging Sixties' so too did skirts, and when economic indicators fell (as during the Great Depression of the thirties, the period of austerity following the Second World War, and the new recession in the seventies) skirt lengths followed suit. Attempts by fashion houses to go against these currents (such as the 'midi-skirt' project) have failed dismally. Of course this could be mere coincidence, but it might also reflect the fact that when times are good people feel free to allow themselves the luxury of play, and accordingly they dress in more sexually provocative ways (rather as they do at parties and holiday resorts).

Although there are marked differences between cultures as regards general levels of permissiveness and tolerance, certain uniformities should not be overlooked. All societies have taboos against rape and incest, for example, even though they may set slightly different criteria for defining acceptable limits. Nearly all societies operate some kind of double standard according to which women (especially wives) are expected to show more sexual reserve and responsibility than men. Not all societies exercise such drastic penalties against female transgression as do the Arabs, but even the most permissive cultures such as our own regard virginity and fidelity as more becoming to a woman than a man. It seems that cultural norms have an in-built allowance for the inevitability of male wanderlust. Women then have been appointed the guardians of moral control and family stability.

The ancient and ubiquitous profession of prostitution represents an interesting dilemma for society. On the one hand, it supports family life and the virtue of the majority of girls in society by siphoning off a great deal of the male need for novelty and stimulation without 'good' girls falling victim. On the other hand, the moral guardians of society find it difficult to tolerate the presence of a hard core of very 'sinful' women in their midst. The result is considerable ambivalence and confusion about whether to forbid prostitution outright or merely minimize its public visibility.

Other aspects of human sexuality that are fairly independent of the cultural setting include recognition of the fact that women need longer to 'warm up' than men before they are able to obtain sexual satisfaction, and the fact that youth and beauty are more highly valued attributes in women than men. In summary, the discovery of cultural differences in many aspects of sexuality should not blind us to the many important uniformities that may be observed, and it certainly does not mean that instinctual tendencies are irrelevant. Cultural idiosyncrasies are superimposed upon very powerful biological imperatives.

Women's liberation

It is often said that gender differences relating to the double standard are retreating fast with the advent of women's liberation. For the last two decades, women have been encouraged by the media to throw off the shackles of sexual repression and to become free, or even assertive, in the bedroom. So it is supposed that modern women are demonstrating that their sexual needs are a great deal more like those of men than has hitherto been realized and that gender differences in sexual inclination are diminishing if not actually disappearing. While this may be true for a small minority of highly sexed women, there is precious little indication that any change is taking place in the population at large.

Last year I conducted a survey of sexual attitudes and behaviour with the help of Britain's best-selling national daily newspaper, *The Sun*. Over four thousand men and women returned questionnaires anonymously by post, and for purposes of analysis these were divided into two broad age-groups of above and below thirty (an arbitrary division consisting of a round number which happened to split the total sample roughly in half.) If women's liberation is having a major impact on modern sexual behaviour, we might expect to find that traditional gender differences would be less pronounced in the younger generation than the older one. In other words, the under thirties in the sample (whose average age was twenty-three) should be more liberated than the over thirties (whose average age was forty). In fact, no such tendency emerged (Wilson, 1980). One of the key questions relating to parental investment theory asked respondents how soon after meeting a new partner they would normally like to have sexual intercourse. About one third of the men wanted sex at the first possible opportunity, compared with only nine per cent of women. Most women preferred to wait until they had some commitment to a steady relationship before having sex; some would even hold out until marriage. The striking thing about the results, however, was the absence of any difference between older and younger respondents in their replies to this question. The modern generation of girls are apparently just as reticent sexually as their mothers.

Another key question concerned the nature of 'ideal sex'. As expected, men expressed more frustration overall than women, and in particular, they would have liked a greater variety of partners than they were enjoying, as well as trying more exciting variations on the sex act. Of those women who did express a desire for more sex, the majority would have liked more straight sex with their present partner or spouse. Again, the consistency between the two age-groups in all of these respects was quite remarkable; the two sets of data might easily have been drawn randomly from the same population. The modern generation of girls, it seems, remain traditionally feminine in that they focus their libido upon one loved partner (while their male partners continue to lust after novelty and excitement). This may be a great disappointment to the ideologists of sex equality but it does appear to be the truth of the matter.

Other interesting results from the *Sun* survey were the proportions of men and women experiencing various sexual difficulties. For all the articles in magazines like *Cosmopolitan* giving women advice on how to achieve orgasm, the girls of today appear to be having just as much trouble as previous generations. Forty-four per cent of women under thirty reported orgasm difficulty, compared with forty-one per cent of the over thirties. Likewise, the problems of guilt, anxiety and disgust associated with sex are no less common in the younger generation than the old, despite an enormous amount of propaganda to the effect that sex is good and wholesome. Perhaps even more intriguing is the discovery that a great many women both young and older, are plagued by simple boredom and lack of interest. This was a much more common complaint among women (being registered by one in every three) than anxiety, and it challenges the assumption of many sex therapists in the Masters and Johnson mould that women usually have sufficient libido if only it can be unleashed. For many women, it appears, libido is not so much blocked by counter forces like fear and guilt; it is simply not strong enough to generate an adequate sex life. Treatment might therefore often be better aimed at trying to raise libido either by hormonal or erotic stimulation, rather than at reducing anxiety. Gillan and Gillan (1976) offer some useful suggestions along these lines.

Some people might question the choice of *The Sun* for a survey of this kind. It does have a very wide readership with an occupational profile that matches that of Britain as a whole, which makes it arguably the most suitable single publication for the purpose. On the other hand, it devotes quite a lot of attention to sexual matters, and this, together with selective response to the questionnaire, means that the sample may well overrepresent people who are especially preoccupied with sex and who are fairly permissive. This, however, should not affect results concerning gender and age differences. In fact, it makes it all the more surprising that so many women should report lack of interest in sex as a problem, and that so much female reserve should appear.

The Sun adorns its pages with bare-breasted women and this might deter some feminists committed to opposing the treatment of women as sex objects. Such women would more likely read *The Guardian*, but they would constitute a small minority of the population, and in any case are not the sort of women who would be expected to promote 'liberated' sexual behaviour on the part of women. There are at least two different points of view concerning the kind of sex equality that is desirable. Some people believe that women should feel free to use pornography, take sexual initiative and enjoy a variety of partners in the same way that men have traditionally done. Others are more like 'new puritans', believing that men should be prevented from seeing pornography and that their insensitive, lustful tendencies should be curtailed as far as possible in favour of responsible, monogamous behaviour. Whether these two attitudes can suitably be identified with the labels 'women's liberationist' and 'feminist' respectively is open to argument.

In noting that modern women have not changed as regards sexual attitudes and behaviour, I do not mean to suggest that the women's movement has been impotent. Considerable advances have been made towards equality in educational, occupational and economic spheres. These are fully justified and are no doubt with us to stay. Women today are also less likely to feel guilty about being atypical of their sex-role stereotype. For example, the minority of women who do have the same sexual proclivities as men will feel more comfortable in the modern social climate about admitting and exercising them.

Such changes have certainly occurred and should be applauded. The movement only becomes undesirable, in my opinion, when scientific evidence is denied because it is politically uncomfortable, and women who would otherwise have been contented wives and mothers are made to feel like inadequate failures because they have not pursued careers outside the home.

The kibbutz experiment

Women's liberation is not actually a new idea. A brave and fascinating experiment was conducted by the Israelis when they set up their rural communes, the kibbutzim, in the early part of this century. A central part of their ideology was the total emancipation of women from all the inequalities (sexual, social, economic, and intellectual) that had been imposed upon them by traditional society. Since they believed that the burden of child-rearing and home-making lay at the root of sex-role differentiation and female inequality, radical changes in family structure were instituted. Traditional marriage was abolished and replaced by a system of cohabitation in which a man and woman were assigned shared sleeping accommodation within the commune but retained their own separate names and identities. Children were reared with their age peers in state nurseries where they played, ate, slept, were educated, and had minimal contact with their parents. Adults were supposed to think of all the kibbutz children as joint social property and were discouraged from developing any special relationship with their own off-spring. Thus freed of the 'domestic yoke' women were persuaded to engage in productive work to the same extent as men and all the symbols of femininity such as female clothes, cosmetics, jewellery and feminine hair styles were rejected. Apparently, it was assumed that, to be the equals of men, women would have to look like men as well as share their occupational roles.

When anthropologists Melford and Audrey Spiro studied the achievements of the kibbutzim in 1950, the experiment seemed to have been fairly successful and their view of human nature as 'culturally relative' was confirmed. However, when in 1975, Melford Spiro returned to the kibbutz for a follow-up study he found that, in the intervening quarter century, dramatic changes had occurred in the domains of marriage, family and sex-role differentiation which all but undid the earlier revolution (Spiro, 1979). The new generation of women, although raised with unisex models (with plenty of women driving tractors and men in domestic service occupations), and having been raised from infancy with propaganda to the effect that men and women are identical by nature, were now pressing to be allowed fulfillment in the mother role. The kibbutz government had become predominantly male (apparently because of female uninterest in politics) and a traditional sex division of labour had become established. Marriage had reverted to its original form, with a wedding ceremony and celebration, and public displays of attachment and 'ownership' that were previously taboo. The units of residence had changed from the group to the married couple, and couples were now claiming the right to enjoy the company of their own children. Children were now sleeping with their own parents and spending much more time with them. The women had also shown a return to femininity in dress and general self-presentation, personality and hobbies.

While this return to a pre-revolutionary form of sex differentiation might be explained in terms of exposure to outside (city) influences, Spiro thought this interpretation very unlikely. In studying play references of the kibbutz children, for example, he found that the girls most often played 'mother' (bestowing care and affection on a doll or small animal, for instance) while the most frequent role adopted by boys, was that of an animal – not the domestic animals with which they were familiar, but wild and ferocious animals like snakes and wolves. No form of social learning theory could explain why girls should adopt a culturally appropriate model (the parenting woman) in their fantasy play, while boys adopt a culturally irrelevant model (wild animals). We are then left with biological tendencies towards nurturance and aggression respectively as an explanation of this difference. A careful examination of evidence such as this led Spiro to conclude that the counter-revolution he had observed in the kibbutz must have represented a return to nature rather than the effect of reactionary social forces. For a man previously committed to 'cultural relativity theory', this was a considerable change of mind.

The first indication of a confrontation between nature and ideology in the kibbutz concerned the issue of public nudity. The pioneers of the movement had early on decided that sexual equality would best be promoted by ignoring all differences in male and female anatomy. Therefore, boys and girls in the children's houses were raised in a totally 'sex-blind' environment. They used the same toilets, dressed in each other's presence, and showered together. This system of unashamed nudity worked perfectly well until the girls reached puberty, at which point (quite spontaneously and in opposition to prevailing social pressures) they developed intense feelings of embarrassment and began to demand privacy. An active rebellion against the mixed showers was initiated, with the girls showering separately and refusing to admit the boys into the showers at the same time. For some time the authorities refused to change the system, but eventually they became convinced that it amounted to a 'form of torture' for the girls, and today nearly all kibbutz high schools have separate shower rooms for boys and girls. Again, if outside cultural influences were responsible for this failure of ideology it is hard to see why they should not have been felt by younger girls or older boys, but should strike selectively at the pubescent girls. The modesty which girls develop at puberty is apparently not a result of social guilt-induction; it is part of the natural coyness which serves the female strategy of selectivity.

Implications

From this brief account of the role of social factors in sexual behaviour we may conclude several things. Firstly, cultural effects are real. We of the human species have it within our power to organize social affairs in many different ways. To acknowledge a powerful biological component to our emotions and desires does not necessarily imply fatalism. We can modify or even override our animal instincts completely, and it will often be morally desirable to do so. In devising a moral code, however, it will help to take out biological nature into consideration so that sources of stress and instability can be identified in advance.

Secondly, cultures differ more in overall levels of permissiveness than they do with respect to instinctual tendencies and sex-role differences. From the more liberal societies, such as our own (Eysenck, 1976), to the more conservative, such as Japan (Wilson and Iwawaki, 1980), a more responsible and selective attitude towards sexual matters is expected by, and of, women than men. The double standard is not imposed upon women by men, it is endorsed as valid by women at least equally often.

Thirdly, the effects of culture upon sex-role differentiation are not one-directional. That is, all social pressures do not operate in such a way as to increase gender differences. We have seen that in the kibbutzim an attempt was made to collapse gender differences by social learning processes – an attempt that eventually proved unsuccessful. Our own society may well have reached a point where the social pressures opposed to sex-role differentiation are more powerful than those which make for an exaggeration of natural differences. It is difficult to say, but there are certainly plenty of conflicting voices that may be heard and alternative models that may be followed. A teenage boy may be prompted by his peers to 'score' with as many girls as possible, but his own parents would be unlikely to recommend such behaviour, and he is bound to have had contradictory messages from other agencies such as the Church. A young girl might gather that her parents expect her to be virginal, but her boy friends would present skillful counter-propaganda, and many voices in the media would be heard to extol the virtues and pleasures of sexual freedom. Social learning theory does not adequately explain why males and females choose to respond to messages which are compatible with mating strategies that seem optimal from an evolutionary point of view.

The reader may wonder why traditional female behaviour (chastity and monogamy) is usually regarded as 'good', while typical male behaviour (promiscuity and polygyny) is classified as morally undesirable. Surely the most moral position would be one that 'split the difference', giving an equal degree of justice (and causing an equal degree of distress) to each sex. One suggested explanation is that since women are entrusted with the early moral education of children they have taken the opportunity to indoctrinate both sexes with their own ethical system. I don't think this is true, however. It is far more likely that we have come to regard female behaviour as morally superior to that of males because it *is* morally superior. As already noted, it is male instincts that create the vast majority of social problems; without female stability and compassion the world would be a great deal more insecure and hostile than it is already. Nevertheless, it does seem unfair that morality is largely female defined; true justice should require that women make some concession to male nature.

In European society we frequently fail to communicate with members of the opposite sex because we underestimate the extent of the differences between men and women. Erin Pizzey, in discussing a woman's lot following divorce, was quoted as follows: 'I was appalled at the number of men who thought a woman on her own would be willing to sleep with anyone, just for the sake of companionship. I preferred the pain of loneliness to that' (*New Standard*, 20 November 1980). This is an interesting statement which reveals considerable honesty on both sides. Miss Pizzey is no prude, but her feelings are very characteristically female. How many men would pretend to be interested in a woman, even in love with her, in order to gain sex? And reciprocally, how many women would use sex (either the giving or withholding) in order to gain long-term devotion (i.e. a husband)? Not only are the male and female mating strategies in conflict with one another, but we are often motivated to deceive the opposite sex that our own inclinations are more like theirs than they really are. Thus we see the hypocrisy of judges and vicars who patronize brothels and yet condemn prostitutes, and women who are provocatively flirtatious at parties but have no real interest in going to bed. The deception involved may often be unintentional, but can easily lead to misunderstandings concerning the nature of the opposite sex.

Another reason why many people underestimate the difference between typical men and women is that they have memorable contact with exceptions to the stereotype. Men are likely to have casual encounters with women who are atypically libidinous and variety seeking, and women are likely to have their deepest relationships with men who are monogamously inclined, at least at the time. It is easy for each sex to go away with the belief that these experiences are more representative than they really are.

The true situation with respect to gender differences is only understood when it is appreciated that the overlap between men and women in mental dispositions is probably about the same as that for physical attributes such as height. The differences in instinct are real, but they apply only on average. It is therefore important that people be judged as individuals, and not assumed always to be typical members of their sex. The kind of society we should seek is one that gives maximum opportunity and freedom for individuals to express themselves regardless of whether or not their inclinations conform to their gender stereotype. This is the just and desirable part of the feminist cause.

By the same token, we need to be careful that women are not deterred from finding fulfillment within a traditional feminine role. The fact of gender differences does not mean that women should be regarded as inferior in any way. Equality does not depend upon identity, and if men and women do gravitate towards different social roles there is no reason to think these are not of equivalent value. As one kibbutz woman pointed out, men's work is mostly concerned with 'things' and animals, while the occupations preferred by women are concerned with people. Which is more important?

I do not write this book with the intention of fomenting the sex war. Rather, it is my hope that a fuller understanding of the instinctual sources of conflict between the sexes will lead to more constructive attempts at communication. We must be honestly aware of what we have to contend with in human nature if we are to make a lasting social peace.

References

Bagley, C., 1969, 'Incest behaviour and incest taboo', *Social Problems*, 16, 505-519.

Bardis, P.D., 1979, 'Homeric love' in M. Cook and G.D. Wilson (Eds), *Love and Attraction* (Oxford: Pergamon).

Bardwick, J.M., 1971, *Psychology of Women: A Study of Biocultural Conflicts* (New York: Harper and Row).

Beach, F., 1975, 'In pursuit of the intellectual orgasm' (Interview with Fleming and Maxey), *Psychology Today*, Vol. 8 No. 10 (March), 69-77.

Bell, A.P., 1974, 'Homosexualities: their range and character', in 1973 *Nebraska Symposium on Motivation* (Lincoln: University of Nebraska Press).

Bermant, G., 1976, 'Sexual behaviour: Hard times with the Coolidge Effect', in M.H. Siegel and H.P. Zeigler (Eds), *Psychological Research: The Inside Story* (New York: Harper and Row).

Bermant, G. and Davidson, J.M., 1974, *Biological Bases of Sexual Behaviour* (New York: Harper and Row).

Birtchnell, J., 1979, 'A test of Toman's theory of mate selection', in M. Cook and G.D. Wilson (Eds), *Love and Attraction* (Oxford: Pergamon).

Blumstein, P.W. and Schwartz, P., 1976, 'Bisexuality: Some social psychological issues', *Proceedings of American Sociological Association* (New York).

Bowlby, J., 1971, *Attachment* (Harmondsworth: Penguin).
Burton, F.D., 1970, 'Sexual climax in female *macaca mulatta*', *Proceedings of the Third International Congress of Primatology* (Zurich).
Carney, A., Bancroft, K. and Matthews, A., 1978, 'Combination of hormonal and psychological treatment for female sexual unresponsiveness. A comparative study', *British Journal of Psychiatry*, 132, 339-346.
Christensen, H.T. and Gregg, C.F., 'Changing sex norms in America and Scandinavia', *Journal of Marriage and the Family*, 32, 616-627.
Daly, M. and Wilson, M., 1979, 'Sex and strategy', *New Scientist*, 4 January.
Darwin, C.R., 1871, *The Descent of Man and Selection in Relation to Sex* (London: Murray. Revised edition 1881).
Davenport, W.H., 1965, 'Sexual patterns and their regulation in a society of the South-West Pacific', in F.A. Beach (Ed.), *Sex and Behaviour* (New York: Wiley).
Demorest, W.J., 1977, 'Incest avoidance among human and non-human primates', in S. Chevalier-Skilnikoff and F.E. Poirier (Eds), *Primate Biosocial Development* (New York: Garland).
Dewhurst, C.J. and Gordon, R.R., 1969, *The Intersexual Disorders* (London: Balliere).
Dobash, R. and Dobash R., 1980, *Violence against wives* (London: Open Books).
Drewett, R.F., 1973, 'Sexual behaviour and sexual motivation in the female rat', *Nature*, 242, 476-477.
Dunbar, R., 1978, 'Life with geladas: A battle of the sexes', *New Scientist*, 6 July, 28-30.
Eibl-Eibesfeldt., I, 1971, *Love and Hate: On the Natural History of Basic Behaviour Patterns* (London: Methuen).
Epstein, A.W., 1961, 'Relationship of fetishism and transvestism to brain and particularly temporal lobe

dysfunction', *Journal of Nervous and Mental Diseases*, 133, 247-253.

Epstein, A.W., 1975, 'The fetish object: Phylogenetic considerations', *Archives of Sexual Behaviour*, 4, 303-308.

Eysenck, H.J., 1976, *Sex and Personality* (London: Open Books).

Eysenck, H.J. and Wilson, G.D., 1979, *The Psychology of Sex* (London: Dent).

Fisher, R.A., 1930, *The Genetical Theory of Natural Selection* (Oxford: Clarendon Press).

Fisher, S., 1973, *Understanding the Female Orgasm* (Harmondsworth: Penguin).

Fox, C.A., Wolff, H. and Baker, J.A., 1970, 'Measurement of intra-vaginal and intra-uterine pressures during human coitus by radio telemetry, *Journal of Reproduction and Fertility*, 22, 243-251.

Giallombardo, R., 1974, *The Social World of Imprisoned Girls* (New York: Wiley).

Gillan, P. and Gillan R., 1976, *Sex Therapy Today* (London: Open Books).

Glass, S.P. and Wright, T.L., 1977, 'The relationship of extramarital sex, length of marriage and sex differences on marital satisfaction and romanticism: Athanasio's data reanalysed', *Journal of Marriage and the Family*, 39, 691-703.

Goldstein, N.J., Kant, H.S., Judd, L.L., Rice, C.J. and Green, R., 1971, 'Exposure to pornography and sexual behaviour in deviant and normal groups', in *Technical Report of the Commission on Obscenity and Pornography. Vol VII* (Washington D.C.: U.S. Govt. Printing Office).

Gosselin C.C. and Wilson, G.D., 1980, *Sexual Variations: Fetishism, sadomasochism and transvestism* (London: Faber and Faber).

Harcourt, A.H. and Stewart, K.J., 1977, 'Apes, sex and

societies', *New Scientist*, 20 October, 160-162.
Harlow, H.F. and Harlow, M.K., 1965, 'The affectional systems', in *Behaviour of Nonhuman Primates*, Vol. 2, A.M. Schrier, H.F. Harlow and F. Stollnitz (Eds) (London: Academic Press).
Hartley, S.F. and Pietraczyk, L.M., 1979, 'Preselecting the sex of offspring: Technologies, attitudes and implications', *Social Biology*, 26, 232-246.
Hatkoff, T.S. and Laswell, T.E., 1979, 'Male-female similarities and differences in conceptualizing love', in M. Cook and G.D. Wilson (Eds), *Love and Attraction* (Oxford: Pergamon).
Hess, E., 1975, *The Tell-tale Eye* (New York: McMillan).
Heston, L.L. and Shields, J., 1968, 'Homosexuality in twins: A family study and a registry study', *Archives of General Psychiatry*, 18, 149-160.
Hite, S., 1976, *The Hite Report* (New York: McMillan).
Hoenig, J. and Kenna, J.C., 1979, 'EEG abnormalities and transsexualism', *British Journal of Psychiatry*, 134, 293-300.
Hunt, M., 1974, *Sexual Behaviour in the 1970's* (Chicago: Playboy Press).
Hutchinson, G.E., 1959, 'A speculative consideration of certain possible forms of sexual selection in man', *American Naturalist*, 93(869), 81-91.
Imperato-McGinley, J., Guerrero, L., Gautier, T. and Peterson, R.E., 1974, 'Steroid 5-alpha-Reductase deficiency in man: An inherited form of male pseudohermaphroditism', *Science*, 186, 1213-1215.
Jedlicka, D., 1980, 'A test of the psychoanalytic theory of mate selection', *Journal of Social Psychology*, 112, 295-99.
Kallmann, F.J., 1952, 'Comparative twin study on the genetic aspects of male homosexuality', *Journal of Nervous and Mental Diseases*, 115, 283-298.

Kane, F.J., Lipton, M.A. and Ewing, J.A., 1969, 'Hormonal influences in female sexual response', *Archives of General Psychiatry*, 20, 202-209.

Kolarsky, A., Freund, K., Machek, J. and Polak, O., 1967, 'Male sexual deviation, association with early temporal lobe damage', *Archives of General Psychiatry*, 17, 735-743.

Krafft-Ebing, R., 1886, *Psychopathia Sexualis* (12th Edition), translated by F.S. Klaf (New York: Stein and Day, new edition 1965).

Lee, J.A., 1976, *Lovestyles* (London: Dent).

Levi, L., 1968, 'Sympatho-adrenomedullary activity, diuresis and emotional reactions during visual sexual stimulation in females and males', *Reports from the Laboratory for Clinical Stress Research* (Depts of Medicine and Psychiatry, Karolinska sjukhuset, Stockholm), August.

Lorenz, K., 1952, *King Solomon's Ring* (London: Macmillan).

Lukianowicz, N., 1972, 'Incest', *British Journal of Psychiatry*, 120, 301.

Marshall, D.S., 1971, 'Sexual behaviour on Mangaia', in D.S. Marshall and R.C. Suggs (Eds), *Human Sexual Behavior* (New York: Basic Books).

Masters, W.H. and Johnson, V.E., 1966, *Human Sexual Response* (Boston: Little, Brown).

Mead, M., 1961, 'Cultural determinants of sexual behaviour', in W.C. Young (Ed.), *Sex and Internal Secretions*, Vol II (Baltimore: Williams and Wilkins).

Mead, M., 1967, *Male and female: A Study of the Sexes in a Changing World* (New York: William Morrow).

Mead, S.L. and Rekers, G.A., 1979, 'Role of the father in normal psychosexual development', *Psychological Reports*, 45, 923-931.

Miller, A.R., 1969, 'Analysis of the Oedipal complex', *Psychological Reports*, 24, 781-782.

Morgan, P., 1975, *Child care: Sense and Fable* (London: Temple Smith).

Morris, D., 1971, *Intimate Behaviour* (London: Cape).

Morris, D., 1977, *Manwatching: A Field Guide to Human Behaviour* (London: Cape).

Nadler, R.D., 1975, 'Sexual cyclicity in captive lowland gorillas', *Science*, 189, 813.

Nadler, R.D., 1977, 'Sexual behaviour of captive orangutans', *Archives of Sexual Behaviour*, 6, 457-475.

Nichols, R.C., 1978, *Heredity and environment: Major findings from twin studies of ability, personality and interests* (Unpublished MS, State University of New York at Buffalo).

Nicholson, J., 1979, *A Question of Sex* (London: Fontana).

Parkes, A.S., 1976, *Patterns of Sexuality and Reproduction* (London: Oxford University Press).

Peplau, L.A., Rubin, Z. and Hill, C.T., 1977, 'Sexual intimacy in dating relationships', *Journal of Social Issues*, 33, 86-109.

Pigache, P., 1978, 'Erotopathology: The aetiology of falling in love', *World Medicine*, 3 May.

Pinkava, V., 1971, 'Logical models of sexual deviations', *International Journal of Man-Machine Studies*, 3, 351-374.

Rachman, S.J. and Hodgson, R.J., 1968, 'Experimentally induced "sexual fetishism" ', *Psychological Record*, 18, 25-27.

Reinisch, J.M., 1977, 'Prenatal exposure of human foetuses to synthetic progestin and oestrogen affects personality', *Nature*, 266, 561-562.

Robertson, D.R., 1972, 'Social control of sex reversal in a coral-reef fish', *Science*, 177, 1007-1009.

Rose, R.M., Bernstein, I.S., and Gordon, T.P., 1975, 'Consequences of social conflict on plasma testosterone levels in rhesus monkeys', *Psychosomatic Medicine*, 37, 50-61.

Rossi, A.S., 1973, 'Maternalism, sexuality, and the new feminism', in J. Zubin and J. Money (Eds), *Contemporary Sexual Behaviour: Critical Issues in the 1970s* (Baltimore: The John Hopkins University Press).

Schlegel, W.S., 1975, 'Parameter Beckenskelett', *Sexualmedizin*, 4, 228-232.

Schofield, M., 1965, *The Sexual Behaviour of Young People* (Harmondsworth: Penguin).

Shelley, D.S.A. and McKew, A., 1979, 'Pupillary dilation as a sexual signal and its links with adolescence', in M. Cook and G.D. Wilson (Eds), *Love and Attraction* (London: Pergamon).

Sherfey, M.D., 1972, *The Nature and Evolution of Female Sexuality* (New York: Random House).

Shope, D.F., 1968, 'The orgastic responsiveness of selected college females', *Journal of Sex Research*, 4, 206-219.

Sigusch, V. and Schmidt, G., 1971, 'Lower class sexuality: some emotional and social aspects in West German males and females', *Archives of Sexual Behaviour*, 1, 29-44.

Simms, T.M., 1967, 'Pupillary response of male and female subjects to pupillary difference in male and female picture stimuli', *Perception and Psychophysics*, 2, 553-555.

Sipova, I. and Starka, L., 1977, 'Plasma testosterone values in transsexual women', *Archives of Sexual Behaviour*, 6, 441-455.

Spiro, M.E., 1979, *Gender and Culture: Kibbutz women revisited* (Durham, NC: Duke University Press).

Stoller, R.J., 1975, *Perversion: The Erotic Form of Hatred* (New York: Delta).

Symons, D., 1979, *The Evolution of Human Sexuality* (New York: Oxford University Press).

Tavris, C.D. and Sadd, S., 1977, *The Redbook Report on Female Sexuality* (New York: Delacorte Press).

Trivers, R.L., 1972, 'Parental investment and sexual selection', in B. Campbell (Ed.), *Sexual Selection and the Descent of Man* (Chicago: Aldine).

Vance, E.B. and Wagner, N.N., 1976, 'Written descriptions of orgasm: A study of sex differences', *Archives of Sexual Behaviour*, 5, 87-98.

Wallin, P., 1960, 'A study of orgasm as a condition of women's enjoyment of intercourse', *Journal of Social Psychology*, 51, 191-198.

Wales, E. and Brewer, B., 1976, 'Graffiti in the 1970s', *Journal of Social Psychology*, 99, 115-123.

Wilson, E.O., 1975, *Sociobiology: The New Synthesis* (Cambridge, Mass.: Belknap).

Wilson, G.D., 1978, *The Secrets of Sexual Fantasy* (London: Dent).

Wilson, G.D., 1981, 'Cross-generational stability of gender differences in sexuality', *Personality and Individual Differences*, 2, in press.

Wilson, G.D. and Fulford, K.W.M., 1979, 'Sexual behaviour, personality and hormonal characteristics of heterosexual, homosexual and bisexual men', in M. Cook and G.D. Wilson, (Eds), *Love and Attraction* (Oxford: Pergamon).

Wilson, G.D. and Iwawaki, S., 1980, 'Social attitudes in Japan', *Journal of Social Psychology*, 112, 175-180.

Wilson, G.D. and Nias, D.K.B., 1976, *Love's Mysteries: The Psychology of Sexual Attraction* (London: Open Books).

Zuger, B., 1979, 'Effeminate behaviour present in boys from childhood: Ten additional years of follow-up', *Comprehensive Psychiatry*, 19, 363-369.